Memory Based Question Bank on Fruit Science

NIPA® GENX ELECTRONIC RESOURCES & SOLUTIONS P. LTD.
New Delhi-110 034

Memory Based Question Bank on Fruit Science

Niranjan Singh

Assistant Professor
Department of Fruit Science
Acharya Narendra Deva University of Agriculture and Technology
Ayodhya (UP), India

D.P. Sharma

Professor and Head
Department of Fruit Science
Dr. Yashwant Singh Parmar University of Horticulture & Forestry
Nauni, Solan, Himachal Pradesh

Neeraj Sankhyan

Assistant Professor
Dr. Yashwant Singh Parmar University of Horticulture & Forestry
Nauni, Solan, Himachal Pradesh

NIPA® GENX ELECTRONIC RESOURCES & SOLUTIONS P. LTD.

New Delhi-110 034

NIPA® GENX ELECTRONIC RESOURCES & SOLUTIONS P. LTD.

101,103, Vikas Surya Plaza, CU Block
L.S.C.Market, Pitam Pura, New Delhi-110 034
Ph : +91 11 27341616, 27341717, 27341718
E-mail:newindiapublishingagency@gmail.com
www: www.nipabooks.com

For customer assistance, please contact
Phone: + 91-11-27 34 17 17
Fax: + 91-11-27 34 16 16

ISBN: 978-81-19235-05-6

NIPA also publishes books in a variety of electronic formats. Some content that appears in print may not be available in electronic books, and vice versa.

Composed and Designed by NIPA.

Preface

Pomological research principally aims at the advancement, improvement, cultivation, and physiological investigation of fruit trees. The objectives of fruit tree augmentation involve the improvement of fruit quality, the regulation of production periods, and the minimization of production expenses.

This publication is beneficial for JRF, SRF, NET, ARS, M.Sc. Entrance, State, and Central Level Competitive Examinations. It is designed to be objective-oriented and user-friendly.

Contents

Preface v

1. ICAR – IARI Ph.D. Fruit Science Entrance Exam – 2010 1
2. ICAR – IARI Ph.D. Fruit Science Entrance Exam – 2011 17
3. ICAR – IARI Ph.D. Fruit Science Entrance Exam – 2013 35
4. ICAR – IARI Ph.D. Fruit Science Entrance Exam – 2015 53
5. ICAR – IARI Ph.D. Fruit Science Entrance Exam – 2016 67
6. ICAR – NET / ARS Fruit Science Exam – 2011 83
7. ICAR – NET / ARS Fruit Science Exam – 2012 97
8. ICAR – NET / ARS Fruit Science Exam – 2013 113
9. ICAR – NET / ARS Fruit Science Exam – 2014 129
10. ICAR – NET / ARS Fruit Science Exam – 2015 145
11. ICAR – NET Fruit Science Exam – 2016 161
12. ICAR – NET Fruit Science Exam – 2017 177
13. ICAR – NET Fruit Science Exam – 2018 193
14. ICAR – NET Fruit Science Exam – 2019 209
15. ICAR – NET Fruit Science Exam – 2020 225
16. ICAR – NET Fruit Science Exam – 2021 233
17. ICAR – NET Fruit Science Exam – 2023 249
18. ICAR – NET Fruit Science Exam (Sample Paper-1) 267
19. ICAR – NET Fruit Science Exam (Sample Paper-2) 281
20. ICAR – NET Fruit Science Exam (Sample Paper-3) 297

1

ICAR – IARI Ph.D. Fruit Science Entrance Exam– 2010

1. The share of agriculture and allied activities in India's GDP at constant prices in 2008-09 was:
   a) 15.1% b) 17.1%
   c) 21.6% d) 24.0%
2. The total area under transgenic crop in India during 2008-09 was:
   a) 4.4 million ha b) 6.4 million ha
   c) 8.4 million ha d) 10.4 million ha
3. Fertilizer consumption in India during 2009-10 was:
   a) 128 kg/ha b) 138 kg/ha
   c) 148 kg/ha d) 158 kg/ha
4. The genome of which of the following crops is still not completely sequenced?
   a) Rice b) Aerabidopsis
   c) Papaya d) Wheat
5. The study of fresh water is known as:
   a) Limnology b) Hydrology
   c) Synecology d) Autecology
6. 'Aadhar' is related to:
   a) Unique identification card b) University identification
   c) FCI buffer stock d) None of these
7. The first self-replicating, synthetic bacterial cell was created by:
   a) Craig Venter b) T.J. Burrill
   c) A.V. Leeuwenhoek d) Louis Pasteur
8. Balanced fertilizer ratio (NPK) recommended for legumes is:
   a) 1:2:1 b) 1:2:2
   c) 4:2:1 d) 3:2:1

9. Which of the following fertilizer is neutral in nature?
   a) MOP b) Urea
   c) CAN d) KCI
10. For applying 50 kg of nitrogen, how much area should one use?
   a) 99 b) 109
   c) 119 d) 129
11. Which of the following is commonly referred to as muriate of potash?
   a) Potassium nitrate b) Potassium chloride
   c) Potassium sulphate d) Potassium silicate
12. Which among the following is another name for vitamin B1?
   a) Niacin b) Pyridoxine
   c) Thiamine d) Riboflavin

b) The GM crops have the highest area in the world in order of preference is
   c) Nutritional enhancement > virus resistant
   d) Virus resistant > herbicide tolerance
   e) Herbicide tolerance > insecticide-resistant
   f) Biofuels > herbicide tolerance

13. Which one of the following is an Indian native fruit crop?
   a) Mango b) Papaya
   c) Sapota d) Grape
14. Virus-free plants can be obtained through:
   a) Anther culture b) Meristem culture
   c) Embryo culture d) Ovule culture
15. ELISA test is conducted for detection of:
   a) Bacteria b) Fungi
   c) Nematode d) Virus
16. Which one the following is an introduced species of the honeybee?
   a) *Apis indica* b) *Apis mellifera*
   c) *Apis florea* d) *Apis dorsata*
17. Which variety of rice is known to have the longest seed length?
   a) Pusa-1121 b) Pusa-2212
   c) Pusa Basmati-1 d) IR-64

18. The blue colour label on the insecticide container represents which degree of toxicity level:
    a) Extremely toxic
    b) Highly toxic
    c) Moderately toxic
    d) Slightly toxic
19. White bud disorder of maize is caused due to the deficiency of:
    a) Zn
    b) Cu
    c) Mn
    d) Fe
20. Which one of the following is not a measure of central tendency?
    a) Mean
    b) Median
    c) Mode
    d) Coefficient of variance
21. Which one of the following is not a measure of dispersion?
    a) Range
    b) Mean deviation
    c) Standard deviation
    d) Mode
22. The causal organism of ear cockle disease of wheat is:
    a) *Anguina tritici*
    b) *Meloidogyne inocognita*
    c) *Ditylenchus angustus*
    d) *Heterodera avenae*
23. The Minimum support price of crops is fixed on the recommendation of:
    a) CACP
    b) GEAC
    c) ICARA
    d) FCI
24. Which rice variety is resistant to stem borer, BLB and blast
    a) IR-24
    b) IR-36
    c) IR-64
    d) Pusa Basmati-I
25. Which of the following herbicide is commonly used to control Phalaris minor in wheat?
    a) 2, 4-D
    b) Isoproturon
    c) Pendimethalin
    d) Atrazine
26. A fast-growing crop planted between two major crops as a contingency crop is known as:
    a) Cash crop
    b) Catch crop
    c) Intercrop
    d) Companion crop
27. Which of the following condition is true for Xi and Yi to have a negative correlation:
    a) Xi increases and Yi decreases
    b) Xi increases and Yi increases
    c) Xi decreases and Yi decreases
    d) Xi decreases and Yi increases

28. The carbohydrate translocation in germinating seeds is due to:
    a) IBA b) GA3
    c) ABA d) Kinetin
29. Which of the following cell organelle is involved in root gravity pool?
    a) Amyloplast b) Mitochondria
    c) Nucleus d) Chloroplast
30. Mule tail / crazy top is a genetic disorder of:
    a) Almond b) Apricot
    c) Cherry d) Walnut
31. 'Verdelli', whichrefers to the summer production of flower following water stress, is associated with which fruit crop?
    a) Guava b) Lemon
    c) Mango d) Pomegranate
32. Sugar Loaf; the sweetest variety of pineapple belongs to which group?
    a) Cayenne group b) Abacaxi group
    c) Spanish group d) Queen group
33. Genotypes of the sex reversing male papaya are:
    a) M_1 M^{RR} and M_1 M^{Rr} b) $M_2 M^{RR}$
    c) $M_1 M_2$ d) mm
35. Mangiferin is a/an:
    a) Anti-malformin b) Malformin
    c) Flowering agent d) None of these
36. Shot berry formation is a major problem in which of the following grape varieties :
    a) Perlette b) Thompson Seedless
    c) Anab-e-Shahi d) Bangalore Blue
37. Which of the following is a multiple hybrid (three-way cross) variety of pomegranate?
    a) Ruby b) Mridula
    c) Bhagwa d) Phule Arkta
38. The fruit which is mostly used as a vegetable is:
    a) Jackfruit b) Lasoda
    c) Kokuma d) Avocado

39. Chilling requirement of walnut is hrs below 7.2°C.

    a) 100-700 b) 700-1400

    c) 1400-2100 d) 2100-2700

40. The causal organism of powdery mildew of ber (Indian jujube) is:

    a) *Isariopsis indica* b) *Alternaria alternata*

    c) *Odium eriosiphoides* d) *Colletotrichum spp.*

41. Which of the following crop is heliotropic?

    a) Coconut b) Cocoa

    c) Banana d) Date palm

42. Which of the following is an indigenous variety of arecanut?

    a) Mangla b) Sumangla

    c) Sreemangla d) Mohitnagar

43. The blue disease of banana is caused due to the deficiency of:

    a) Mn b) Mg

    c) Fe d) Zn

44. 'Albinism' is a physiological disorder of:

    a) Banana b) Pineapple

    c) Strawberry d) Papaya

45. Which of the following rootstocks is recommended for HDP of Kinnow mandarin?

    a) Carrizo citrange b) Rangpur lime

    c) Troyer citrange d) Karna khatta

46. Which of the following rootstocksincreases ascorbic acid (vitamin C) content in guava?

    a) *Psidium acutangulum* b) *Psidium cattleianum*

    c) *Psidium pumilum* d) *Psidium cujavillus*

47. Windbreak for protecting orchards from strong winds is planted on which direction of the orchard?

    a) South-West b) North-East

    c) South-East d) North-West

48. Which of the following is the best pollinizer for Dushehari mango?

    a) Langra b) Bombay Green

    c) Chausa d) Kesar

49. Which of the following is a non-astringent cultivar of persimmon?
    a) Jiro b) Hachiya
    c) Triumph d) Tanenashi
50. Firing disorder is associated with:
    a) Guava b) Apple
    c) Apricot d) Pear
51. Which chemical is used for breaking bud dormancy in grape?
    a) Hydrogen cyanamide b) KNO3
    c) Hydrogen peroxide d) All of these
52. Polyploid variety of grape, grown commercially in Germany is:
    a) Perle b) Kyoho
    c) Pione d) Olympia
53. In mango,......................has been recommended as inter-stock for Amrapali grafted on Mallika.
    a) Anupam b) Totapuri Red small
    c) Creeping d) Vellaicollumban
54. A variety of litchi suitable for canning is:
    a) Rose scented b) Bombai
    c) Shahi d) Swarna Roopa
55. Which of the following is a late-maturing cultivar of loquat?
    a) Saharanpur Special b) Thames Yellow
    c) Fire Wall d) Tanaka
56. Golden Yellow is a self-incompatible variety of:
    a) Cherry b) Walnut
    c) Loquat d) Apple
57. Konkan Amrit is a variety of:
    a) Mangosteen b) Mahua
    c) Kokum d) Pomegranate
58. Zill, a Floridan mango cultivar, is a selection of:
    a) Mulgoa b) Haden
    c) Eldon d) Irwin
59. Optimum temperature for mango cultivation is:
    a) 10-14°C b) 14-18°C
    c) 18-22°C d) 24-27°C

60. Mariana, an important rootstock of plum, has been developed from a cross between:
    a) Prunus cerasifera × Prunus munsoniana
    b) Prunus cerasifera × Prunus armeniaca
    c) Prunus cerasifera × Prunus munsoniana
    d) *Prunus cerasifera* × *Prunus simonii*
61. Which of the following is the sweetest variety of mango?
    a) Alphonso b) Kesar
    c) Dushehari d) Chausa
62. Best combiner variety used in the hybridization of mango is:
    a) Bangalora b) Neelum
    c) Alphonso d) Dushehari
63. Insitu method of mango grafting is:
    a) Epicotyl grafting b) Inarching
    c) Softwood grafting d) Veneer grafting
64. Kinnow mandarin, the first generation hybrid between King orange (*Citrus nobilis*) and Willow leaf mandarin (*Citrus deliciosa*) was developed by:
    a) H.B. Frost b) W.T. Swingle
    c) T.Tanaka d) Spigeal Roy
65. Best irrigation method for the newly planted orchard is:
    a) Drip irrigation b) Basin irrigation
    c) Floor irrigation d) Sprinkler irrigation
66. Titron Early is a .... chill plum variety.
    a) Low chill b) High chill
    c) Moderate chill d) Zero chill
67. Which of the following is a micronutrient loving plant?
    a) Banana b) Citrus
    c) Grape d) Papaya
68. Which variety of banana is used for making baby food?
    a) Kunnan b) Nendran
    c) Monthan d) Hill banana
69. Bee activity drastically reduces below temperature.
    a) 15°C b) 20°C
    c) 25°C d) 30°C

70. Black nose in date palm occurs during which of the following developmental stage?
    a) Gandora stage b) Doka stage
    c) Pind stage d) Dang stage
71. Which variety of grape is commercially cultivated in North India?
    a) Thompson Seedless b) Perlette
    c) Anab-e-Shahi d) Balck Champa
72. Which fruit crop is resistant to Mn deficiency?
    a) Walnut b) Peach
    c) Pecan nut d) Strawberry
73. Best time for pruning in grape in north India is:
    a) October-November d) December-January
    c) March-April d) June-July
74. The papain having the activity of .... tyrosine units is most preferred for trading purpose.
    a) 400-600 b) 600-800
    c) 800-1000 d) 1000-1200
75. Which of the following Ziziphus species is a host for lac insect?
    a) *Ziziphus horsifieldi* b) *Ziziphus oenoplia*
    c) *Ziziphus xylopyrus* d) *Ziziphus incurva*
76. The best rootstock of ber for budding of many varieties is:
    a) *Ziziphus nummularia* b) *Ziziphus rotundifolia*
    c) *Ziziphus rugosa* d) *Ziziphus incurva*
77. Which of the following is/are unusual variety/varieties of mango?
    a) Chitla Afaq b) Croton
    c) Dofasla d) All of these
78. The average percentage of hermaphrodite flowers in mango is:
    a) 20% b) 40%
    c) 60% d) 80%
79. Which of the following is a drought-tolerant rootstock of apple?
    a) M 9 b) MM 106
    c) MM 111 d) MM 109

80. One-year-old shoot of grape which bears fruits is known as:
    a) Bud b) Cane
    c) Tendril d) Trunk
81. Peel cracking is a maturity index of:
    a) Walnut b) Litchi
    c) Cherry d) Bael
82. NBPGR, New Delhi has established a gene sanctuary for citrus in:
    a) Garo hills of Meghalaya b) Khasi hills of Meghalaya
    c) Nilgiri hills of Tamil Nadu d) Darjeeling hills of West Bengal
83. National Horticulture Mission (NHM) was launched in the year:
    a) 2004-05 b) 2005-06
    c) 2006-07 d) 2007-08
84. the total seed count in apple is:
    a) 4 b) 6
    c) 8 d) 10
85. Phyllanthoid branching habit is a characteristic feature of:
    a) Aonla b) Avocado
    c) Phalsa d) Karonda
86. The first order sylleptic branches are usually formed in:
    a) Zizyphus mauritiana b) Emblica officinalis
    c) Aegle marmelos d) Grewia subinaequalis
87. Flowering in aonla occurs on which type of shoots?
    a) Determinate b) Indeterminate
    c) Apical d) All of the above
88. April pruning of grapes under Maharashtrian condition is known as:
    a) Foundation pruning b) Forward pruning
    c) Fruit pruning d) None of these
89. The red colour of grapefruit is due to the presence of:
    a) Anthocyanin b) Lycopene
    c) Carotene d) Quercetin
90. Which of the following varieties of mango is suitable for processing?
    a) Amrapali b) Mallika
    c) Pusa Surya d) Pusa Arunima

91. Type of inflorescence in mango is:
    a) Solitary b) Spadix
    c) Panicle d) Catkin
92. Higher phloem to xylem ratio in mango is an indication of:
    a) Dwarfness b) Regularity in bearing
    c) Disease resistant d) Early maturity
93. 'Calyptra' is a term associated with which of the following fruit crops?
    a) Litchi b) Grape
    c) Quince d) Apricot
94. Characteristic of Vitis parviflora is/ are:
    a) Reflexed stamen b) Early maturity
    c) Anthracnose resistant d) All of these
95. Pusa Arunima; a mango hybrid is a cross between:
    a) Dushehari x Sensation b) Amrapali x Sensation
    c) Alphonso x Sensation d) Amrapali x Neelum
96. Ploidy level of Pusa Srijan is:
    a) 2n+1 b) 2n+2
    c) 2n-1 d) 2n-2
97. Lethal stem pitting is a disease of:
    a) Loquat b) Avocado
    c) Carambola d) Olive
98. The total number of sex forms in papaya as proposed by Story are:
    a) 3 b) 8
    c) 32 d) 58
99. Fruit type of Wood apple is:
    a) Syconus b) Amphisaraca
    c) Berry d) Drupe
100. Which European country grows mangoes commercially?
    a) Germany b) France
    c) Italy d) Spain
101. The botanical name of cucumber trees is:
    a) Averrhoa bilimbi b) Averrhoa carambola
    c) Averrhoa microphylla d) Averrhoa minima

102. The best planting material for pineapple is:

a) Slips b) Suckers

c) Crown d) Stolen

103. Which of the following is an ultra-dwarfing rootstock of citrus?

a) Citrus limonia b) Citrus karna

c) Troyer citrange d) Flying Dragon

104. Which of the following banana varieties has apple-like flavour?

a) FHIA-1 b) FHIA-2

c) FHIA-3 d) FHIA-4

105. Genomic constitution of Rasthali is:

a) AAA b) AAB

c) ABB d) AB

106. Which of the following varieties of ber is resistant to fruit fly?

a) Gola b) Umran

c) Seb d) Tikdi

107. One of the oldest fruit crops known to mankind is:

a) Date palm b) Fig

c) Grape d) Olive

108. Jonathan spot of apple is caused due to the deficiency of :

a) Ca b) B

c) Moisture d) Zn

109. Hen and Chicken, a disorder of grapes is caused due to the deficiency of:

a) Ca b) Zn

c) B d) Fe

110. Dichotomous branching habit is found in:

a) Cashew nut b) Loquat

c) Phalsa d) Karonda

111. Cherry grafted onto colt rootstock is planted at a height of:

a) 3 X 3 m b) 5 X 5 m

c) 7 X 7 m d) 10 X 10 m

112. Spring fruiting comes on one-year-old shoot in:

a) Mandarin b) Guava

c) Grape d) Pomegranate

113. *Phyllanthus acidus* is commonly known as:

a) Star gooseberry b) Otaheite gooseberry

c) Country gooseberry d) All of these

114. Pramalini variety of acid lime was released from:

a) MPKV, Rahuri b) KKV, Dapoli

c) MAU, Parbhani d) PDKV, Akola

115. Which of the following is a sister seedling of Perlette?

a) Beauty Seedless b) Delight

c) Flame Seedless d) Himrod

116. The major problem in the breeding of pineapple is:

a) Self-sterility b) Self-incompatibility

c) Heterostyly d) All of these

117. Highest productivity of grape has been reported from:

a) China b) Italy

c) USA d) India

118. Which of the following is an auto-tetraploid variety of banana?

a) Bodles Altafort b) Klue Teparod

c) Saba d) FHIA-1

119. Apetalous flowers are found in:

a) Litchi b) Mango

c) Grape d) Ber

120. Avocado is recommended as a high energy food for diabetic patients because it is a rich source :

a) Poly-unsaturated fatty acids b) Mono-unsaturated fatty acids

c) Protein d) Minerals

121. Optimum pH for orchard planting is:

a) 5.5-6.5 b) 6.5-7.5

c) 7.5-8.5 d) 8.5-9.5

122. Time of FBD in mango under north Indian condition is:

a) February-March b) July-August

c) October-November d) December-January

123. Which of the following is the most dwarfing variety of papaya?

a)	Pusa Dwarf	b)	Pusa Nanha
c)	Pusa Majesty	d)	Pusa Delicious

124. Which of the following is a low chilling variety of apple?

a)	Fenny	b)	Ambri
c)	Winter Banana	d)	Red Delicious

125. Which of the following is/are a variety/varieties of tamarind?

a)	PKM 1	b)	Urigam
c)	Prathisthan	d)	All of these

126. Aonla variety NA-4 is a seedling selection from:

a)	Banarasi	b)	Francis
c)	Chakaiya	d)	BSR-1

127. Cherry variety suitable for canning is:

a)	Early Richmond	b)	North Star
c)	Montmorency	d)	English Morello

128. Fruit with persistent calyx is:

a)	Litchi	b)	Mangosteen
c)	Mahua	d)	None of the above

129. Pulp colour in guava is controlled monogenically. Which one is correct about the inheritance of red pulp colour in guava?

a) Red pulp colour is recessive
b) Red pulp colour is dominant over white
c) Red and white colour are co-dominant
d) None of the above

130. Phytophthora can be effectively controlled by using bio-agent:

a)	*Trichoderma harzianum*	b)	*Aspergillus niger*
c)	*Pseudomonas fluorescens*	d)	*Bacillus subtillis*

131. Match the Mango cultivars with their respective characteristics:

Cultivar		Characteristic	
i)	Langra	a)	Inferior quality fruits
ii)	Alphonso	b)	Most prone to fruit drop
iii)	Fazli	c)	Susceptible to bacterial leaf spot
iv)	Amrapali	d)	Limited adaptability
v)	Bangalora	e)	Uneven fruit size

132. Match the following pests and diseases with their corresponding Biocontrol agents:

Pests and diseases		Biocontrol agent	
i)	Woolly apple aphid	a)	*Brunus saturalis*
ii)	Sanjose scale	b)	*Platymeris laevicollis*
iii)	Coconut rhinoceros beetle	c)	*Cryptolaemus montrouzieri*
iv)	Citrus mealy but	d)	*Encarsia pernicious*
v)	Citrus whiteflies	e)	Aphelinus mali

133. Match the given grape varieties with their corresponding physiological disorders:

Variety		Disorder	
i)	Perlette	a)	Compact cluster
ii)	Beauty Seedless	b)	Post-harvest berry drop
iii)	Anab-e-Shahi	c)	Shot berry
iv)	Thompson Seedless	d)	Pink berry
v)	Bangalore Blue	e)	Uneven ripening

134. Match the following diseases and their causal organisms:

Disease		Causal organisms	
i)	Cigar end rot of banana	a)	*Taphrina deformans*
ii)	Pierce's disease of grape	b)	*lllCerotelium fici*
iii)	Peach leaf curl	c)	*Liberibacter asiaticus*
iv)	Fig rust	d)	*Verticillium theobromae*
v)	Citrus greening	e)	*Xylella fastidiosa*

135. Match the following Genes and their role in plants:

Gene		Role in plants	
i)	Chitnase	a)	Male sterility
ii)	Coat protein	b)	Virus resistance
iii)	Burmese	c)	Fungal resistance
iv)	Attacin E	d)	Plant architecture
v)	Rol c gene	e)	Bacterial resistance

136. Match the following Citrus rootstocks and their characteristic:

Rootstock		Characteristic	
i)	Cleopatra mandarin	a)	Drought tolerant
ii)	Flying Dragon	b)	Ultra dwarfing
iii)	*Severinia buxifolia*	c)	Salt tolerant
iv)	Carrizo citrange	d)	Cold hardy
v)	Rangpur lime	e)	Nematode resistant

137. Match the following physiological disorders and causes:

Disorder		Cause	
i)	Bud killing	a)	Excessive N2
ii)	Interveinal chlorosis	b)	Mg deficiency
iii)	Leaf scorching	c)	B deficiency
iv)	Blossom end rot	d)	Ca deficiency
v)	Fruit cracking	e)	Cl toxicity

138. Match the following fruit crops with the corresponding major cultural practices:

Fruit crop		Related term	
i)	Banana	a)	Caprification
ii)	Mango	b)	Cincturing
iii)	Litchi	c)	Girdling
iv)	Grape	d)	Mattocking
v)	Fig	e)	Smudging

139. Match the following families with their corresponding genre

Family		Genus	
i)	Ebenaceae	a)	Ribes
ii)	Clusiaceae	b)	Durian
iii)	Proteaceae	c)	Diospyros
iv)	Saxifragaceae	d)	Garcinia
v)	Bombaceae	e)	Macadamia

140. Match the following fruits with their corresponding common names

i)	Guava	a)	Apple of carthage
ii)	Banana	b)	Sugar apple
iii)	Pomegranate	c)	Apple of tropics
iv)	Custard apple	d)	Apple of paradise
v)	Ber	e)	Apple of desert

Answers Key

1.	(b)	2.	(c)	3.	(a)	4.	(d)	5.	(a)	6.	(a)	7.	(a)
8.	(b)	9.	(c)	10.	(b)	11.	(b)	12.	(c)	13.	(c)	14.	(a)
15.	(b)	16.	(d)	17.	(b)	18.	(a)	19.	(c)	20.	(a)	21.	(d)
22.	(d)	23.	(a)	24.	(a)	25.	(b)	26.	(b)	27.	(a)	28.	(a)
29.	(b)	30.	(a)	31.	(a)	32.	(b)	33.	(b)	34.	(a)	35.	(b)
36.	(a)	37.	(a)	38.	(a)	39.	(b)	40.	(c)	41.	(a)	42.	(d)
43.	(b)	44.	(c)	45.	(c)	46.	(d)	47.	(d)	48.	(b)	49.	(a)
50.	(d)	51.	(a)	52.	(a)	53.	(a)	54.	(c)	55.	(d)	56.	(c)
57.	(c)	58.	(b)	59.	(d)	60.	(a)	61.	(d)	62.	(b)	63.	(c)
64.	(a)	65.	(b)	66.	(a)	67.	(b)	68.	(a)	69.	(a)	70.	(b)
71.	(b)	72.	(d)	73.	(b)	74.	(c)	75.	(c)	76.	(b)	77.	(d)
78.	(b)	79.	(c)	80.	(b)	81.	(a)	82.	(a)	83.	(b)	84.	(d)
85.	(a)	86.	(a)	87.	(a)	88.	(a)	89.	(b)	90.	(a)	91.	(c)
92.	(a)	93.	(b)	94.	(d)	95.	(b)	96.	(b)	97.	(b)	98.	(c)
99.	(b)	100.	(d)	101.	(a)	102.	(a)	103.	(d)	104.	(a)	105.	(b)
106.	(d)	107.	(b)	108.	(a)	109.	(c)	110.	(d)	111.	(b)	112.	(d)
113.	(d)	114.	(c)	115.	(b)	116.	(b)	117.	(d)	118.	(a)	119.	(a)
120.	(b)	121.	(b)	122.	(c)	123.	(b)	124.	(c)	125.	(d)	126.	(c)
127.	(c)	128.	(b)	129.	(b)	130.	(a)						

Q	131.	132.	133.	134.	135.	136.	137.	138.	139.	140.
i)	(b)	(e)	(c)	(d)	(c)	(d)	(a)	(d)	(c)	(c)
ii)	(d)	(d)	(a)	(e)	(b)	(b)	(b)	(e)	(d)	(d)
iii)	(c)	(b)	(b)	(a)	(a)	(c)	(e)	(b)	(e)	(a)
iv)	(e)	(c)	(d)	(b)	(e)	(e)	(d)	(c)	(a)	(b)
v)	(a)	(a)	(e)	(c)	(d)	(a)	(c)	(a)	(b)	(e)

2

ICAR – IARI Ph.D.
Fruit Science Entrance Exam – 2011

1. Which of the following crops have been approved for commercial cultivation in India?

   a) Bt cotton and Bt brinjal
   b) Bt cotton and Golden Rice
   c) Bt maize and Bt. cotton
   d) Bt cotton only

2. The expected food grain production in India for the year 2010-11 was:

   a) 212 million tonnes
   b) 220 million tonnes
   c) 235 million tonnes
   d) 250 million tones

3. The genome of which of the following crops is still not completely sequenced?

   a) Rice
   b) Soybean
   c) Sorghum
   d) Wheat

4. According to the Approach Paper to the 12th Five year Plan, the basic objective of the 12th plan is:

   a) Inclusive growth
   b) Sustainable growth
   c) Faster, more inclusive and sustainable growth
   d) Inclusive and sustainable growth

5. To address the problems of sustainable and holistic development of rainfed areas including appropriate farming and livelihood system approaches, the Government of India has set up the:

   a) National Rainfed Area Authority
   b) National Watershed Development Project for Rainfed Areas
   c) National Mission on Rainfed Areas
   d) Command Area Development and Water Management Authority

6. Which of the following sub-schemes are not covered under the Rashtriya Krishi Vikas Yojana?

   a) Extending the Green Revolution to eastern India

   b) Development of 60,000 pulses and oilseeds villages in identified watersheds

   c) National Mission on Saffron

   d) National Mission on Bamboo

7. The minimum support price for the common variety of paddy announced by the Government of India for the year 2010-11 was:

   a) 1030 b) 1000

   c) 980 d) 950

8. According to the Human Development Report 2010 of the United Nations, India's rank in terms of the human development index is:

   a) 119 b) 134

   c) 169 d) 182

9. Which of the following does not apply to SRI method of paddy cultivation?

   a) Reduced water application

   b) Reduced plant density

   c) Increased application of chemical fertilizers

   d) Reduced age of seedlings

10. Which organic acid, often used as a preservative, occurs naturally in cranberries, prunes, cinnamon and cloves?

    a) Citric acid b) Benzoic acid

    c) Tartaric acid d) Lactic acid

11. Cotton belongs to .... family:

    a) Cruciferae b) Anacardiaceae

    c) Malvaceae d) Solanaceae

12. Photoperiodism is:

    a) Bending of shoot towards source of light

    b) Effect of tight/dark durations on physiological processes

    c) Movement of chloroplast in cell in response to light

    d) Effect of light on chlorophyll synthesis

13. Ergot disease is caused by which pathogen on which host?
    a) Claviceps purpurea on rye b) Puccinia recondite on wheat
    c) Drechlera sorokiniana on wheat d) Albugo Candida on mustard

14. Rocks are the chief sources of parent materials over which soils are developed. Granite, an important rock, is classified as:
    a) Igneous rock b) Metamorphic rock
    c) Sedimentary rock d) Hybrid rock

15. Which one of the following is a Kharif crop?
    a) Pearl millet b) Lentil
    c) Mustard d) Wheat

16. The coefficient of variation (C.V.) is calculated by the formula:
    a) (Mean/S.D.) x 100 b) (S.D./Mean) x 100
    c) S.D./Mean d) Mean/S.D.

17. Which of the following is commonly referred to as muriate of potash?
    a) Potassium nitrate b) Potassium chloride
    c) Potassium sulphate d) Potassium silicate

18. Inbred lines that have the same genetic constitution but differ only at one locus are called:
    a) Multi lines b) Monohybrid
    c) Isogenic lines d) Pure lines

19. For applying 100kg of nitrogen, how much urea should one use?
    a) 45 kg b) 111 kg
    c) 222 kg d) 333 kg

20. The devastating impact of plant disease on human lives was first realized by epidemic of:
    a) Brown spot of rice in Bengal b) Late blight of potato in USA
    c) Late blight of potato in Europe d) Rust of wheat in India

21. The species of rice (Oryza) other than *O. sativa* that is cultivated is:
    a) *O. rufipugon* b) *O. Longisteminata*
    c) *O. glaberrima* d) *O. nivara*

22. The enzyme responsible for the fixation of CO_2 in mesophyll cells of C-4 plants is:
    a) Malic enzyme
    b) Phosphoenol pyruvate carboxylase
    c) Phosphoenol pyruvate carboxykinase
    d) RuBP carboxylase

23. Which one of the following is a 'Vertisol'?
    a) Black cotton soil
    b) Red sandy loam soil
    c) Sandy loam sodic soil
    d) Submontane (Tarai) soil
24. What is the most visible physical characteristic of cells in metaphase?
    a) Elongated chromosomes
    b) Nucleus visible but chromosomes not
    c) Fragile double stranded loose chromosomes
    d) Condensed paired chromosomes on the cell plate
25. All weather phenomena like rain, fog and mist occur in..........layer:
    a) Troposhere
    b) Mesophere
    c) Ionosphere
    d) Ozonosphere
26. Which of the following elements is common to all proteins and nucleic acids?
    a) Sulphur
    b) Magnesium
    c) Nitrogen
    d) Phosphorous
27. Silt has intermediate characteristics between:
    a) Sand and loam
    b) Clay and loam
    c) Loam and gravel
    d) Sand and clay
28. Certified seed is produced from:
    a) Nucleus seed
    b) Breeder seed
    c) Foundation seed
    d) Truthful seed
29. Seedless banana is an:
    a) Autotriploid
    b) Autotetraploid
    c) Allotriploid
    d) Allotetraploid
30. Which one of the following is used to test the goodness-of-fit of a distribution?
    a) Normal test
    b) t-test
    c) Chi-square test
    d) F-test
31. According to Kostermans and Bompard (1993), the number of species of uncertain origin in genus Mangifera is:
    a) 9
    b) 11
    c) 13
    d) 15

32. The first mango crop was introduced in America from:
   a) Egypt b) India
   c) Cuba d) Brazil

33. Under Maharashtrian conditions, fruit bud-differentiation in grape takes place in the month of:
   a) February b) March
   c) April d) May

34. Fruitfulness in Thompson Seedless vines at light intensity of 3,600 ft candles at 35°C is required for:
   a) 12 h b) 13 h
   c) 16 h d) 18 h

35. Grape rootstock 1613 is a hybrid of:
   a) V. champini x Othelo b) V. solonis x Othelo
   c) V. solonis x V. champini d) Othelo x V. soloni

36. Which of the following genotypes is used as donor for nematode resistance in peach?
   a) Prunus ussuriansis b) Prunus davidiana
   c) Pyrus pashla d) Prunus grifithii

37. .... is the donor parent for cold hardiness in pear:
   a) Pyrus pashia b) Pyrus khasiana
   c) Prunus ussuriansis d) Prunus davidiana

38. In banana, the highest productivity has been reported as:
   a) 13 t/ha b) 23 t/ha
   c) 43 t/ha d) 63 t/ha

39. Moroccan datepalm cultivars have been shown to have somatic chromosome number:
   a) 24 b) 26
   c) 32 d) 36

40. Bayoud disease in datepalm is caused due to:
   a) Fusarium b) Pythium
   c) Sclerotina d) Phytophthora

41. Tetraploids in pineapple bear fruits which are:
   a) Extra large b) Large
   c) Small d) Elongated

42. Black heart in pineapple appears when exposed to temperature:

a) <43°C b) <20°C

c) <25°C d) >25°C

43. Which of the following Mangifera species has only tetramerous flowers?

a) *M. altissima* b) *M. pentandra*

c) *M. grifithii* d) *M. laurina*

44. Which of the following mango cultivar is monoembryonic?

a) Kur b) Madu

c) Florigon d) Ono

45. .... is a fibreless Annona species:

a) *A. glabra* b) *A. muricata*

c) *A. montana* d) *A. cherimola*

46. Which of the following species belongs to Muscadinia type?

a) Vitis sylvestris b) Vitis popenoeii

c) Vitis Champini d) Vitis vulpine

47. Grape Fan leaf virus is spread by:

a) Whitefly b) Aphid

c) Bird d) Nematode

48. 'Chicken tongue' refers to the seedless fruit of:

a) Litchi b) Avocado

c) Mangosteen d) Durian

49. Bergamot is derived from the maternal cross:

a) Citrus auraptium x Citrus medica

b) Citrus aurantifolia x Citrus medica

c) Citrus medica x Citrus aurantium

d) Citrus medica x citrus aurantifolia

50. Which of the following is not a valid species of citrus?

a) Citrus reticulata b) Citrus maxima

c) Citrus medica d) Citrus latipes

51. Which of the citrus genus is used as a dye:

a) Citropsis b) Citrus maxima

c) Citrus medica d) Atalantia

52. Which of the following is the native of Australia?
    a) Microcitrus b) Eremocitrus
    c) Clymenia d) Atalantia

53. The constant factor for estimating heat unit is:
    a) 7.5 b) 10.0
    c) 12.5 d) 17.5

54. San Jose scale of apple was introduced in India from:
    a) UK b) USA
    c) Canada d) France

55. M 27 rootstock was evolved from cross between:
    a) M13 x M7 b) M13 x M9
    c) M7 x M13 d) M9 x M13

56. Cork spot in pear appears due to deficiency of:
    a) Ca b) B
    c) Zn d) K

57. Self-incompatibility in plum is governed by:
    a) Single gene b) Oligo gene
    c) Polygene d) Nuclear gene

58. The rootstock Paja of cherry is botanically known as:
    a) Prunus gudouni b) Prunus fruticosa
    c) Prunus puddum d) Prunus cerasoides

59. An ideal fruit crop for producing edible vaccine through genetic transformation:
    a) Red banana b) Pineapple
    c) Papaya d) Tree tomato

60. Which of the following fruits has maximum life on tree before ripening?
    a) Sapota b) Aonla
    c) Avocado d) Mamey Sapota

61. Which of the following is the third generation Floridan mango?
    a) Haden b) Irwin
    c) Zill d) Ruby

62. Apart from Mulgoa, which is the other parent of the Floridan mango varieties?
    a) Carabao b) Turpentine
    c) Haden d) Hindi

63. Which of the following strawberry varieties is known to be day neutral?
    a) Toiga b) Sweet Charlie
    c) Selva d) Pajaro

64. Dichogamy is a rule in walnut but it is absent in:
    a) Namdan b) Sulaiman
    c) Govind d) Payne

65. Jaboticaba is the member of .... family:
    a) Moraceae b) Euphorbiaceae
    c) Myrtaceae d) Mridhula

66. The highest nut yielding cashew variety in India:
    a) Ullal 2 b) Vengurla 2
    c) Vengurla 5 d) Mridhula

67. Baby coconut is commercially known as:
    a) Cocojal b) Coco-nana
    c) Coco-nino d) Coco-Cream

68. Sister seedling of Arka Kanchan grape:
    a) Arka Trishna b) Arka Soma
    c) Arka Krishna d) Arkavati

69. Which of the following is considered as satellite species?
    a) Key lime b) Tahiti lime
    c) C. latipes d) C. paradise

70. Novaria – a variety of banana has been developed using:
    a) Polypoidy b) In vitro mutagenesis
    c) Hybridization d) Embryo rescue

71. Which of the following conditions of pollution is noted in chestnuts:
    a) Protogyny b) Dichogamy
    c) Duodichogamy d) Protandry

72. In almonds, self-fertility is known to be:
    a) Recessive b) Dominant
    c) Polygenic d) Lethal

73. The group of plants in which chlorophyll a: b ratio is highest:
    a) C b) C
    c) CAM d) Xerophyte

74. Which of the following do not have three different sex forms?
    a) Grape b) Datepalm
    c) Papaya d) Strawberry

75. What is the critical temperature for flowering in pear?
    a) 4 C b) 6 C
    c) 8 C d) 12 C

76. Parthenocarpic fruits in pear lack:
    a) Colour b) Smoothness
    c) Size d) Flavour

77. Exposure of plants to which of the following colour lights leads to the enhancement of gibberellic acid?
    a) Red b) Far-red
    c) Blue d) Uv

78. Photoblastic seed germination is observed in:
    a) Raspberry b) Blue berry
    c) Walnut d) Peach

79. The deficiency symptoms of sulphur first appears on:
    a) Young leaves b) Older leaves
    c) Shoot tip d) Young fruits

80. Excess application of Zn leads to imbalance of:
    a) Diploid b) Tetraploid
    c) Octaploid d) Decaplod

81. Indian strawberry *Fragaria vesca* L. has the ploidy level as:
    a) Xylem blockage b) Phloem blockage
    c) Foot rot d) Little leaf

83. Which of the following fruit crops is best suited for acidic soil?
    a) Jamun b) Citrus
    c) Strawberry d) Phalsa
84. Espalier system of training is similar to:
    a) Cordon b) Telephone
    c) Y system d) Kniffin
85. Green berries in grapes can be minimized with the application of:
    a) Eithrel b) Cycocel
    c) NAA d) p-NOA
86. Northern Spy apple is highly susceptible to:
    a) Water core b) Russeting
    c) Bitter pit d) Wooly aphis
87. Apple Chaubattia Agrim has been evolved as:
    a) Selection b) Mutant
    c) Hybrid d) Triploid
88. 'Endoxerosis' in citrus is caused due to ....disorder:
    a) Nutrition b) Moisture
    c) Both (A) and (B) d) Zn
89. Apart from grape, early rain is also harmful for the cultivation of:
    a) Strawberry b) Pomegranate
    c) Lemon d) Datepalm
90. Protein content in cashew nut is:
    a) 22% b) 24%
    c) 26% d) 28%
91. Which of the following is a nocturnal pest of citrus?
    a) Citrus psylla b) Fruit sucking moth
    c) Aphids d) Leaf miner
92. Ficus glomerata rootstock in fig is tolerant to:
    a) Acidic soil b) Sailnity
    c) Nematode d) Fe-toxicity
93. Which Rosaceous fruit crop is rich in pectin:
    a) Peach b) Pear
    c) Quince d) Apricot

94. Feijoa sellowiana was introduced in India from:

a)	Mexico	b)	Brazil
c)	Haiti	d)	West Indies

95. Who is better known as 'Father of Systematic Pomology'?

a)	De Candolle	b)	Drawin
c)	Lamarck	d)	Vavilov

96. The volatile compound present in banana after ripening:

a)	Allypropyl	b)	Linalool
c)	Diallylpropyl	d)	Isopentanol

97. Largest producer of grape fruit:

a)	Spain	b)	USA
c)	Japan	d)	China

98. International Horticulture Congress-2010 of ISHS was held at:

a)	Madrid	b)	Libson
c)	Kent	d)	Paris

99. National Centre for Quality Control of Tissue Culture Raised Plant Material is located at:

a)	IIHR, Banglore	b)	IARI, New Delhi
c)	NRC Citrus, Nagpur	d)	CPRI, Shimla

100. Bizazare orange – a graft chimera has been evolved as:

a)	Sour orange and citron	b)	Sweet orange and citron
c)	Sweet orange and sour orange	d)	Sour orange and grape fruit

101. Which of the following is a coloured spot of apple?

a)	Vareel	b)	Liberty
c)	Jonagold	d)	Supreme

102. Denavelling is done in:

a)	Papaya	b)	Banana
c)	Apple	d)	Pineapple

103. Among temperate fruits, pruning intensity is most severe in:

a)	Apple	b)	Pear
c)	Peach	d)	Plum

104. Tall banana varieties take how many months for harvesting from the time of flowering?

a) 6-7 b) 4-5

c) 3-4 d) 1-2

105. Genomic constitution of FHIA-1 is:

a) AAAA b) ABBB

c) AAAB d) ABB

106. Moko disease of banana is caused due to the attack of:

a) Fungus b) Bacteria

c) Virus d) MLOs

107. The parents of Tangelo citrus are:

a) Mandarin x Sweet orange b) Mandarin x lemon

c) Mandarin x grape fruit d) Mandarin x Sour orange

108. Genus Psidium contains about:

a) 50 spp. b) 60 spp.

c) 80 spp. d) 150 spp.

109. Food Product Order (FPO) was passed in the year:

a) 1950 b) 1953

c) 1955 d) 1960

110. Trichoderma viride is used to control:

a) Thrips b) Powdery mildew

c) Damping off d) Blight

111. The largest number of cultivated vegetables belong to ...... family:

a) Cruciferae b) Malvaceae

c) Fabaceae d) Cucurbitaceae

112. Genetic male sterility is common in:

a) Rose b) Marigold

c) Gladiolus d) Dahlia

113. Ponds and Bridges are some important features of which garden?

a) English b) Mughul

c) Japanese d) Italian

114. How much TPS (True Potato Seed) is required for planting an area measuring one hectare?

a) 20-25 g b) 40-45 g

c) 70-75 g d) 100-150 g

115. Bitterness in almond kernel is governed by:

a) Single dominant gene b) Single recessive gene

c) Mutagens d) Multiple genes

116. Which apple rootstock is resistant to fire blight infection?

a) Robusta-5 b) Irish Peach

c) Northern Spy d) Malus prunifolia

117. Rind of citrus fruits becomes coarser and thicker due to deficiency of which element?

a) Nitrogen b) Zinc

c) Phosphorus d) Magnesium

118. Ellagic acid, a naturally occurring phenolic constituent is found in which of the following fruit crops?

a) Strawberry b) Jackfruit

c) Walnut d) Macadamia

119. Nomilin , which inhibits the development of certain forms of cancer, is present in which fruit crop ?

a) Grapefruit b) Mango

c) Grape d) Ber

120. Profiche is related to spring crop of which fruit crop?

a) Fig b) Longan

c) Persimmon d) Avocado

121. Albedo in citrus fruit is rich in:

a) Protein b) Oil

c) Cellulose d) Avocado

122. Sand pear is scientifically is called:

a) Pyrus Serotina b) Pyrus Communis

c) Pyrus pyrifoliia d) Pyrus cerrasifera

123. Tip layering is a natural method of propagation in which crop?

a) Dewberry b) Raspberry

c) Strawberry d) None of the above

124. Pear decline is related to which of the following?

a)	Pear psylla	b)	Nutrients deficiency
c)	Pathogen attach	d)	All of the above

125. Which is the flavouring ingredient of cherry juice?

a)	Methyl salicylate	b)	Keracyanin
c)	Cyanidin	d)	None of the above

126. Which fruit is the richest source of vitamin B6?

a)	Almond	b)	Guava
c)	Blue berries	d)	Walnut

127. Stooling method of guava multiplication was first reported in which year?

a)	1945	b)	1954
c)	1958	d)	1984

128. Generic name 'cocos' as well as popular name coconut is derived from the Spanish word 'Coco' which means:

a)	Monkey Tongue	b)	Monkey Face
c)	Monkey Hand	d)	None of the above

129. Browning of cut pear fruit is caused due to:

a)	Polyphenolase	b)	Tannins
c)	Phosphorous	d)	None of the above

130. Which of the following plum rootstocks is drought tolerant?

a)	Myrobalan 2-7	b)	Myrobalan 5-2
c)	Marianna 2624	d)	None of the above

131. Match the following names with the corresponding no. of taxanomic species w.r.t. Citrus classification:

Authorities		No. taxonomic species	
i)	Hodgson	a)	> 100
ii)	Tanaka	b)	16
iii)	Singh and Nath	c)	17
iv)	Swingle	d)	36
v)	Bhattacharya and dutta	e)	31

132. Match the following pests and diseases with their corresponding Biocontrol agents:

Pest and diseases		Biocontrol agent	
i)	Citrus psylla	a)	*Metarrhizium anisopliae*
ii)	Apple-European red mite	b)	*Bacillus amyloliguifaci*
iii)	Coconut rhinoceros beetle	c)	*Tamarixia radiate*
iv)	Bud rot of coconut	d)	*Aspergillus niger*
v)	Guava wilt	e)	*Typholodromous* spp.

133. Match following superior selections in grapes with the corresponding scientist's name:

Selection		Scientist	
i)	Sonaka	a)	A.G. Meher
ii)	Manik Chaman	b)	L. Venkataratnam
iii)	Monika	c)	Nana Saheb Kale
iv)	Dilkhush	d)	R.S. Kadlag
v)	Rao Sahebi	e)	T.R. Dabade

134. Match the following diseases and their corresponding causal organisms:

Disease		Causal organisms	
i)	Aonla ring rust	a)	*Elslnoe* sp.
ii)	Ber-black leaf spot	b)	*Phaeophlospora* sp.
iii)	Grape-anthracnose	c)	*Ravonelis* sp.
iv)	Sapota – leaf spot	d)	*Phytophthora* sp.
v)	Cocoa – black pod	e)	*Isoriopsis* sp.

135. Match the following fruits with their corresponding chromosome number:

i)	Pineapple	a)	18
ii)	Dateplam	b)	50
iii)	Aonla	c)	16
iv)	Apple	d)	34
v)	Almond	e)	36

136. Match the following fruits with their corresponding varieties:

i)	Jackfruit	a)	Dwarapudi
ii)	Passion fruit	b)	Singapore
iii)	Sapota	c)	Fire ball
iv)	Loquat	d)	Flavorcot
v)	Apricot	e)	Trellising

137. Match the following fruit crops with their corresponding related terms:

i)	Fruit crop Peach	a)	Related term Bitter pit
ii)	Almond	b)	J.H. Hale
iii)	Plum	c)	Colt
iv)	Apple	d)	Pearless
v)	Cherry	e)	President

138. Match the following fruit crops with their places of origin:

Fruit crop		Place of origin	
i)	Citrus paradisi	a)	East Asia
ii)	Citrus medica	b)	West indies
iii)	Citrus limon	c)	China
iv)	Citrus sinensis	d)	South East Asia
v)	Citrus grandis	e)	India

139. Match the following families with their corresponding scientific names:

Family		Scientific name	
i)	Fabaceae	a)	*Physalis peruviana* L.
ii)	Combretaceae	b)	*Persea americana* M.
iii)	Solanaceae	c)	*Terminalia catappa* L.
iv)	Lauraceae	d)	*Nephelium iappaceum* L.
v)	Spindaceae	e)	*Tamarindus indica* L.

140. Match the following fruits with their corresponding common names:

i)	Guava	a)	Elephant apple
ii)	Banana	b)	Star apple
iii)	Wood apple	c)	Apple of tropics
iv)	Phalsa	d)	Apple of paradise
v)	Jamun	e)	Black plum

Answers Key

1.	(d)	2.	(d)	3.	(d)	4.	(c)	5.	(a)	6.	(d)	7.	(a)
8.	(b)	9.	(c)	10.	(b)	11.	(c)	12.	(b)	13.	(a)	14.	(a)
15.	(a)	16.	(b)	17.	(b)	18.	(c)	19.	(c)	20.	(a)	21.	(c)
22.	(b)	23.	(a)	24.	(d)	25.	(a)	26.	(c)	27.	(d)	28.	(c)
29.	(a)	30.	(c)	31.	(b)	32.	(c)	33.	(d)	34.	(c)	35.	(b)
36.	(b)	37.	(c)	38.	(d)	39.	(b)	40.	(a)	41.	(c)	42.	(b)
43.	(a)	44.	(a)	45.	(b)	46.	(b)	47.	(d)	48.	(a)	49.	(a)
50.	(d)	51.	(d)	52.	(b)	53.	(b)	54.	(d)	55.	(d)	56.	(a)
57.	(c)	58.	(d)	59.	(a)	60.	(d)	61.	(b)	62.	(b)	63.	(c)
64.	(d)	65.	(c)	66.	(c)	67.	(c)	68.	(b)	69.	(d)	70.	(b)
71.	(c)	72.	(a)	73.	(b)	74.	(b)	75.	(c)	76.	(d)	77.	(b)
78.	(b)	79.	(a)	80.	(b)	81.	(a)	82.	(b)	83.	(c)	84.	(d)
85.	(a)	86.	(c)	87.	(b)	88.	(b)	89.	(d)	90.	(a)	91.	(b)
92.	(c)	93.	(c)	94.	(b)	95.	(a)	96.	(d)	97.	(a)	98.	(b)
99.	(b)	100.	(a)	101.	(d)	102.	(b)	103.	(c)	104.	(a)	105.	(c)
106.	(b)	107.	(c)	108.	(d)	109.	(c)	110.	(c)	111.	(d)	112.	(b)
113.	(c)	114.	(d)	115.	(b)	116.	(a)	117.	(c)	118.	(a)	119.	(a)
120.	(a)	121.	(c)	122.	(c)	123.	(a)	124.	(a)	125.	(a)	126.	(d)
127.	(d)	128.	(b)	129.	(a)	130.	(b)						

Q.	131.	132.	133.	134.	135.	136.	137.	138.	139.	140.
i)	(d)	(c)	(c)	(c)	(b)	(b)	(b)	(b)	(e)	(c)
ii)	(a)	(e)	(e)	(e)	(e)	(e)	(d)	(e)	(c)	(d)
iii)	(e)	(a)	(a)	(a)	(a)	(a)	(e)	(a)	(a)	(a)
iv)	(b)	(b)	(b)	(b)	(d)	(c)	(a)	(c)	(b)	(b)
v)	(c)	(d)	(d)	(d)	(c)	(d)	(c)	(d)	(d)	(e)

3

ICAR – IARI Ph.D. Fruit Science Entrance Exam – 2013

1. Who is the present Chairman of Protection of Plant Varieties and Farmers' Right Authority (PPV&ERA)?
   a) Dr. R.R. Hanchinal b) Dr. P.L. Gautam
   c) Dr. S. Nagarajan d) Dr. Swapan K. Datta
2. Which among the following is another name of vitamin B12?
   a) Niacin b) Pyridoxal phosphate
   c) Cobalamin d) Riboflavin
3. Which of the following had the largest share in India's farm export earnings in the year 2011-12:
   a) Basmati rice b) Non-basmati rice
   c) Sugar d) Guar gum
4. The National Bureau of Agriculturally Important Insects was established by ICAR in …., and was earlier known as …….
   a) Bangalore; PDBC
   b) New Delhi; National Pusa Collection
   c) Ranchi; Indian Lac Research Institute
   d) New Delhi; NCIPM
5. The most important sucking pests of cotton and rice are respectively:
   a) *Nilaparvata lugens* and *Aphis gossypii*
   b) *Aphis gossypii* and *Thrips oryzae*
   c) *Amrasca bigultula* and *Scirtothrips dorsalis*
   d) *Thrips gossypii* and *Orseolia oryzae*
6. Which of the following microorganisms causes fatal poisoning in canned fruits and vegetables?
   a) *Aspergillus flavus* b) *Penicillium digitatum*
   c) *Clostridium botulinum* d) *Rhizoctonia saloni*

7. The cause of the great Bengal Famine was:
   a) Blast of rice
   b) Brown spot of rice
   c) Rust of wheat
   d) Karnal bunt of wheat
8. Actinomycetes belongs to:
   a) Fungi
   b) Eukaryote
   c) Mycelia sterilia
   d) None of the above
9. A virus-free clone from a virus infected plant can be obtained by:
   a) Cotyledonary leaf culture
   b) Axenic culture
   c) Stem culture
   d) Meristem tip culture
10. Which of the following is not an objective of the National Food Security Mission?
    a) Sustainable increase in production of rice, wheat and pulses
    b) Restoring soil fertility and productivity at individual farm level
    c) Promoting use of bio-pesticides and organic fertilizers
    d) Creation of employment opportunities
11. Agmarknet, a portal for the dissemination of agricultural marketing information, is a joint endeavour of:
    a) DMI and NIC
    b) DMI and Ministry of Agriculture
    c) NIC and Ministry of Agriculture
    d) DMI and Directorate of Economics and Statistics
12. The share of agriculture and allied activities in India's GDP at constant prices in 2011-12 was:
    a) 14.1%
    b) 14.7%
    c) 15.6%
    d) 17.0%
13. The average size of land holding in India according to Agricultural Census 2005-06 is:
    a) 0.38 ha
    b) 1.23 ha
    c) 1.49 ha
    d) 1.70 ha
14. 'Farmers First' concept was proposed by:
    a) Paul Leagans
    b) Neils Rolling
    c) Robert Chamber
    d) Indira Gandhi

15. In the year 2012, GM crops were cultivated in an area of:
    a) 150 million hectare in 18 countries
    b) 170 million hectare in 28 countries
    c) 200 million hectare in 18 countries
    d) 1.70 million hectare in 18 countries
16. The broad-spectrum systematic herbicide, glyphosate, kills the weeds by inhibiting the biosynthesis of:
    a) Phenylalnine b) Alanine
    c) Glutamine d) Cysteine
17. At harvest, the above ground straw (leaf, sheath and stem) weight and grain weight of paddy crop are 5.5 and 4.5 tonnes per hectare, respectively. What is the harvest index of paddy?
    a) 45% b) 50%
    c) 55% d) 100%
18. Crossingover between non-sister chromatids of homologous chromosomes takes place during;
    a) Leptotene b) Pachytene
    c) Diplotene d) Zygotene
19. The term 'Heterosis' was coined by:
    a) G.H. Shull b) W. Bateson
    c) T.H. Morgan d) E.M. East
20. When a transgenic plant is crossed with a non-transgenic, what would be the zygosity status of the F1 plant?
    a) Homozygous b) Heterozygous
    c) Hemizygous d) Nullizygous
21. The highest per capita consumption of flowers in the world is in:
    a) USA b) India
    c) Switzerland d) Netherlands
22. Which of the following is a very rich source of betalain pigment?
    a) Radish b) Beet root
    c) Carrot d) Red cabbage
23. Dog ridge is:
    a) Salt tolerant rootstock of mango
    b) Salt tolerant rootstock of guava
    c) Salt tolerant rootstock of grape
    d) Salt tolerant rootstock of citrus

24. Which of the following micronutrients is most widely deficient in Indian soils?

a) Zinc and boron b) Zinc and iron

c) Zinc and manganese d) Zinc and copper

25. Which of the following fertilizers is not produced in India?

a) DAP b) Urea

c) Muriate of potash d) TSP

26. What is the estimated extent of salt affected soils in India?

a) 5.42 mha b) 7.42 mha

c) 11.42 mha d) 17.42 mha

27. Which of the following is not a feature of watershed?

a) Hydrological unit b) Biophysical unit

c) Socio-economic unit d) Production unit

28. Correlation between

a) 0 and 1 b) -1 and 1

c) -1 and 0 d) 0 and œ

29. For the data 1, -2, 4, geometric mean is:

a) 2 b) 4

c) -7/3 d) -2

30. The relationship between Arithmetic mean (A), Harmonic mean (H) and Geometric mean (G) is:

a) G2=AH b) G=/A+H

c) H2-GA d) A2=GH

31. In pome fruit species, short shoots are named 'Dards' when they are:

a) Strictly vegetative b) Strictly floral

c) Usually vegetative d) Usually floral

32. The sex form of the kiwifruit cultivar 'Hayward' is:

a) Staminate b) Pistillate

c) Hermaphrodite d) Andromonoecious

33. The concept of effective pollination period (EPP) was developed by:

a) Janick b) Williams

c) Frost d) Swingle

34. 1-Methylcyclopropene (1-MCP) is a competitive inhibitor of:
    a) Ethylene b) ABA
    c) Jasmonic acid d) Salicylic acid
35. The most noticeable impact of climate change on mango trees in India is:
    a) Heavy fruit drop
    b) Erratic flowering
    c) Incidence of mango malformation
    d) Bark splitting
36. Which among the following botanical families is also known as the 'Poison ivy' family?
    a) Juglandaceae b) Rosaceae
    c) Anacardiaceae d) Punicaceae
37. The florigenic promoter, which determines flowering in mango, is synthesized in:
    a) Roots b) Apical buds
    c) Flower buds d) Leaves
38. In context of Australian mango germplasm, the term 'Commons' refers to:
    a) Introductions from South East Asia
    b) Frequently used parents in hybridization
    c) Kensington Pride and its clones
    d) Native mango germplasm
39. The top grape producing country in world is:
    a) China b) Italy
    c) United States d) Spain
40. Agri-Export Zones identified for pineapple are located in the states of:
    a) Tripura and West Bengal b) Tripura and Bihar
    c) Tripura and Sikkim d) Tripura and Andhra Pradesh
41. 'Flying Dragon', a dwarfing rootstock of citrus, is susceptible to:
    a) Citrus tristeza b) *Phytophthora* root rot
    c) Citrus nematode d) Iron chlorosis
42. "Bangkok Golden Apple" is a variety of:
    a) Guava b) Pineapple
    c) Avocado d) Pomegranate

43. International Institute of Tropical Agriculture is located in:
    a) Kenya
    b) Ethiopia
    c) Nigeria
    d) Uganda
44. The somatic chromosome number (2n) of Illaichi cultivar of her *Ziziphus mauritiana* is:
    a) 12
    b) 24
    c) 48
    d) 96
45. Richmond - Lang effect describes one of the basic properties of which plant growth regulator:
    a) Jasmonic acid
    b) Cytokinin
    c) Ethylene
    d) Abscisic acid
46. 'Chicken-tongue' seed formation in litchi is due to:
    a) Parthenogenesis
    b) Stenospermocarpy
    c) Failure of pollination
    d) Nutritional deficiency
47. The plant hormone playing a major role in the dormancy of fertilized ovary of aonla (*Emblica officinalis*) is:
    a) Ethylene
    b) Auxin
    c) ABA
    d) Gibberellin
48. Trifactor hypothesis, based on hormonal regulation, seeks to explain the flowering behaviour of:
    a) Papaya
    b) Litchi
    c) Mango
    d) Peach
49. "Fan leaf degeneration", a viral disease of grapes caused by grapevine fan leaf virus (GFLV), is transmitted by:
    a) Flea beetle
    b) Thrips
    c) Nematode
    d) Phylloxera
50. Which of the following sapota cultivar can produce both round and oval shaped fruits on the same tree?
    a) Cricket Ball
    b) Kalipatti
    c) Oval
    d) Bhuripatti
51. The main photosynthetic product in apple is:
    a) Sorbitol
    b) Mannitol
    c) Fructose
    d) Sucrose

52. Light-induced reduction in the photosynthetic capacity of a plant is referred to as:

    a) Photolimitation b) Photodestruction

    c) Photoinhibition d) Photodegradation

53. Deformed papaya fruits with bumpy skin surfaces are produced due to the deficiency of:

    a) Boron b) Calcium

    c) Manganese d) Potassium

54. What do you understand by "Fuji stain"?

    a) A storage disorder of Fuji apples b) A new clone of Fuji apples

    c) A promising somaclone of Fuji d) An apple product

55. The life line tree of Thar Desert is:

    a) *Phoenix dactylifera* b) *Zizyphus mauritiana*

    c) *Prosopis cineraria* d) *Withania somnifera*

56. Which fruit is known as the 'Grape of the Desert'?

    a) *Salvadora oleoides* b) *Cordia myxa*

    c) *Capparis deciduas* d) *Opuntia ficus*

57. The first order sylleptic branches are usually formed in:

    a) *Ziziphus mauritiana* b) *Emblica officinalis*

    c) Aegle marmelos d) *Grewia subinaequalis*

58. The ideal host for expression of Hepatitis 'B' surface antigen (HBsAg) is:

    a) Banana b) Mango

    c) Papaya d) Guava

59. Taxon with only cultivated representation is known as:

    a) Culton b) Indigen

    c) Landrace d) Cultigen

60. The most suitable plant for bio-fencing is:

    a) Ker b) Karonda

    c) Phalsa d) Khijri

61. Which fruit is known as 'Cuddapah almond'?

    a) Almond b) Chironji

    c) Pistachio nut d) Filbert

62. Which flower is known as the 'Queen of East'?
    a) Dahlia b) Gladiolus
    c) Lilium d) Chrysanthemum
63. Chilling sensitive apples are stored at:
    a) -l to 0°C b) 1 to 2°C
    c) 3 to 5°C d) 8 to 10°C
64. Which pathogen is associated with 'Replant problem' in apple?
    a) *Phytophthora sylvaticum* b) *Phytophthora infestans*
    c) *Phytophthora parasitica* d) *Phytophthora aphanidermatum*
65. Which mango variety has recently become popular in the eastern states of India?
    a) Rataul b) Tommy Atkins
    c) Amrapali d) Pusa Arunima
66. Flattening of branches in sapota is caused by:
    a) Fungus b) Bacteria
    c) Phytoplasma d) Spiroplasma
67. Codling moth is a serious problem of apple in:
    a) Kinnaur (Himachal Pradesh) b) Kullu (Himachal Pradesh)
    c) Leh (Jammu & Kashmir) d) Srinagar (Jammu & Kashmir)
68. For harvesting Delicious apple at right maturity, the Starch Pattern Index (SPI) should be:
    a) 2.5/10 b) 3.5/10
    c) 4.5/10 d) 6.5/10
69. Which of the following is known as 'self- peeling banana'?
    a) *Musa laterita* b) *Musa ingens*
    c) Musa basjoo d) *Musa velutina*
70. Which of the following grape varieties has strong Muscat flavour?
    a) Thompson Seedless b) Bhokri
    c) Cardinal d) Pusa Seedless
71. In epicotyl grafting of mango, the age of rootstock is:
    a) 30-35 days b) 20-25 days
    c) 10-15 days d) 5-6 days

72. Wind break is effective up to a distance of ……times the height:

a) 2	b) 3
c) 4	d) 5

73. 'Suppressed climacteric' varieties are found in:

a) Apple	b) Banana
c) Plum	d) Peach

74. The moisture content of 'desiccated coconut' should be:

a) Less than 1%	b) Less than 3%
c) Less than 5%	d) Less than 6.5%

75. In which fruit crop were the bio-control agents used for the first time for controlling post-harvest diseases?

a) Mango	b) Strawberry
c) Banana	d) Apple

76. The moisture content of dried cashew nuts should be:

a) Nearly 2%	b) Nearly 5%
c) Nearly 8%	d) Nearly 12%

77. Which mango variety from Pakistan posed competition to Indian Alphonso in US markets during 2011?

a) Langra	b) Chaunsa
c) Rataul	d) Aman Dashehari

78. The bergamot orange is a hybrid of:

a) *C. jambhiri* × *C. limon*	b) *C. volkameriana* × *C. sinensis*
c) *C. aurantium* × *C. medica*	d) *C. aurantifolia* × *C. macrophylla*

79. Which of the following acts as a Nodal Agency to coordinate the efforts of Agri-Export Zones (AEZ) on the part of Central Government?

a) APEDA	b) NABARD
c) ICAR	d) NHB

80. The Ministry of Agriculture, Govt. of India had announced the year 2012 as the year of:

a) Floriculture	b) Olericulture
c) Horticulture	d) Post-harvest management

81. A rare species of *Mangifera*, which is endemic to the Philippines is:

a) *Mangifera merrillii*	b) *Mangifera minutifolia*
c) *Mangifera laurina*	d) *Mangifera paludosa*

82. Ricey tissue disorder is found in:

a) Banana b) Apple

c) Guava d) Mango

83. The only indigenous variety of apple grown in India that has extraordinary keeping quality:

a) Ambri b) Sunehari

c) Ambrika d) Swarnima

84. Yellow colour in onion is due to ....pigment:

a) Carotene b) Quercetin

c) Lycopene d) Anthocyanin

85. The range of correlation coefficient lies between:

a) 0 to 1 b) -2 to +2

c) -1 to +1 d) 0 to 100

86. Which of the following Japanese varieties has the largest berry size?

a) Pione b) Kyoho

c) Olympia d) Black Olympia

87. Which of the following grape varieties has non-acylated anthocyanin?

a) Black Prince b) Pinot Noir

c) Chardonnay d) Tempranillo

88. Which of the following metal ion is associated with shift in colour pigments in grape?

a) Al b) Mn

c) Mg d) K

89. Seeds, which can tolerate loss of moisture and whose longevity can be increased by storing at low temperature are called as:

a) Recalcitrant seeds b) Orthodox seeds

c) Pure seeds d) Free seeds

90. True representative variety of *Vitis labrusca* is:

a) Bangalore Blue b) Concord

c) Cheema Sahebi d) Thompson Seedless

91. Black sapota is botanically known as:

a) *Bassia longifolia* b) *Diospyros diagram*

c) *Manilkara zapota* d) *Achras zapota*

92. Doubling of chromosomes may lead to:
    a) Temporary suppression of self-incompatibility
    b) Elimination of self-incompatibility
    c) Embryo sac degeneration
    d) Increase in pollen fertility

93. Which group of cherimoya has acidic pulp with pineapple flavour?
    a) Finger-printed b) Tuberculated
    c) Mammilate d) Umbonate

94. 'Coulere' is a problem of:
    a) Mango b) Mangosteen
    c) Grape d) Banana

95. Critical day and night temperatures, which inhibit anthocyanin formation in coloured grapes are:
    a) 28/26°C b) 32/28°C
    c) 35/30°C d) 37/32°C

96. Anthocyanin pigment in apple fruit is found in the form of:
    a) Lycopene b) Cynaidin-3-glucoside
    c) p-carotene (Xanthophyll) d) 5-dehydroshikimic acid

97. In tamarind, the leaflets fold after dark. This is attributed to:
    a) Free tartaric acid b) Free fatty acid
    c) Lupeol d) ADF bound nitrogen

98. Under which chapter of EXIM Policy 2001, has the concept of Agri-Export Zones (AEZ) been introduced by Govt. of India?
    a) 15th b) 16th
    c) 17th d) 18th

99. Daisy Seedless is a new release in:
    a) Mandarin b) Mango
    c) Strawberry d) Kiwifruit

100. Development of a branch with some intervening period of rest of lateral meristem and the branches is referred toes:
    a) Prolepsis b) Syllepsis
    c) Solasesis d) Meollepsis

101. The tolerant rootstock for Fusarium wilt in *Passiflora* spp. is:

a) *P. giberti* b) *P. dulcis*

c) *P. mollissima* d) *P. manicata*

102. The stage of inflorescence emergence in pineapple is known as:

a) Sorose b) Piping

c) Rosette d) Red heart

103. The first commercial self-fertile cultivar of sweet cherry is:

a) Lapin b) Sweetheart

c) Comfort d) Stella

104. In brinjal, the species which show resistance against *Phomopsis* blight and drought are:

a) *S. khasianum* and *S. sisymbriifolium*

b) *S. macrosperma* and *S. integrifolium*

c) *S. melongena* and *S. torvum*

d) *S. incanum* and *S. incanum*

105. 'Acid growth hypothesis' explains the mechanism of action of:

a) Auxins b) Gibberellins

c) Cytokinin d) Ethylene

106. 'Grape phylloxera' a worldwide problem of grapevine, is basically a/an:

a) Fungus b) Virus

c) Nematode d) Insect

107. A classical example of delayed incompatibility, ghe "black line" symptom in English walnut grafted onto rootstocks such as Paradox, , is caused by:

a) Bacteria b) Virus

c) Phytoplasma d) Protozoa

108. In India, double pruning and double cropping systems in grape arc followed in:

a) Temperate region b) Hot tropical areas

c) Mild tropical areas d) Sub-tropical region

109. National Phytotron Facility is located at:

a) Hyderabad b) Chandigarh

c) New Delhi d) Chennai

110. Which fruit crop is placed in the oxycocus section of the genus Vaccinium?

a) Raspberry b) Strawberry

c) Cranberry d) Goji berry

111. In mango hybridization programme, 'Bangalore' is not considered as a suitable parent because the hybrid possesses:

a) Prominent beak with poor quality fruit

b) Prominent beak with excellent fruit quality

c) Problem of spongy tissue

d) Susceptibility to malformation and anthracnose

112. Flowering in aonla occurs on which type of shoots?

a) Determinate b) Indeterminate

c) Apical d) All of the above

113. The major problem in pbalsft cultivation is considered to be:

a) Non-synchronized ripening b) Susceptibility to foliar diseases

c) Self-incompatibility d) Male sterility

114. Critical leaf/fruit ratio at which flower bud is not formed in Delicious apple:

a) Below b) Below 10

c) Below 25 d) Below 30

115. Cauliflorous type of fruit bearing is found in:

a) Jackfruit b) Custard apple

c) Her d) Apple

116. Dehaulming in potato is done:

a) Before planting b) At the time of planting

c) Before harvesting d) After harvesting

117. When used as a female parent in the hybridization programme, this variety of grape does not need emasculation:

a) Hur b) Gold

c) Rubi Red d) Anab-e-Shahi

118. An example of monoaxial fruit plant is:

a) Pomegranate b) Cherry

c) Phalsa d) Papaya

119. Which one of the following fruits was introduced in India for soil conservation purpose but later, attained the status of a commercial fruit crop?

a) Cashew nut b) Sapota

c) Date palm d) Areca nut

120. For an experiment in mango, a scientist took four doses of fertilizers; two methods of application with five replications. Suggest an appropriate experimental design for this experiment.

a) Complete randomized block design

b) Latin square design

c) Factorial randomized block design

d) Split plot design

121. 'Albinism', a physiological disorder of fruit is generally observed in:

a) Grape b) Strawberry

c) Plum d) Guava

122. The Bombay Green variety of mango is also popularly known as:

a) Sylhet b) Sarauli

c) Rataul d) Tikari

123. Mulgoa variety of India was introduced in Florida during:

a) 1879 b) 1889

c) 1899 d) 1909

124. A grape variety resistant to powdery mildew:

a) Regale b) Delaware

c) Chardonnay d) Nistur

125. Enology is the study of:

a) Monument b) Wines

c) Drying of horticultural produce d) *In-vitro* propagation

126. Photosynthetically active radiation (PAR) wave length range is:

a) 0.1 to 0.3 μm b) 0.4 to 0.7 μm

c) 0.7 to 1.0 μm d) 1.0 to 1.3 μm

127. Significance of difference among several means is tested with the help of:

a) x2 test b) t-test

c) F-test d) Z-test

128. A physiological disorder called 'Dry Neck' is common in:

a)	Sapota	b)	Mangosteen
c)	Tree tomato	d)	Avocado

129. Genome constitution of Safed Velchi banana is:

a)	AA	b)	AAA
c)	ABB	d)	AB

130. Hyper sensitive reaction indicates:

a)	High disease susceptibility	b)	True resistance
c)	Tolerance	d)	Escape

131. Match the fruit crops with their physiological disorders:

Fruit crop		Physiological disorders	
i)	Avocado	a)	Premature fruit softening
ii)	Sweet Cherry	b)	Fruit drop
iii)	Banana	c)	Mesocarp discolouration
iv)	Sapota	d)	Rain cracking
v)	Persimmon	e)	Blue disease

132. Match the micronutrients with their corresponding function:

Micronutrient		Function	
i)	Boron	a)	Chlorophyll synthesis
ii)	Zinc	b)	Urea hydrolysis
iii)	Manganese	c)	Pollen germination
iv)	Nickel	d)	Plastocyanin synthesis
v)	Copper	e)	Auxin synthesis

133. Match the names of Scientists with their corresponding contributions:

Scientist		Contribution	
i)	Jude Grosser	a)	Somatic embryogenesis
ii)	Uri Lavi	b)	Transgenic papaya
iii)	Silviero	c)	Somatic Sansavini hybridization
iv)	D. Gonsalves	d)	Mango breeding
v)	R.E. Litz	e)	Fruit physiology

134. Match the following fruits with their respective synonym:

Fruit		Synonym	
i)	Cordia myxa	a)	Christ's thorn
ii)	Garcinia	b)	Alligator pear mangosteen
iii)	Carissa	c)	Cherry of carandas desert
iv)	Averrhoa carambola	d)	Queen of fruits
v)	*Persea americana*	e)	Star fruit

135. Match the given fruit crops with their respective Chromosome number:

Fruit crop	Chromosome number
i) Manilla tamarind	a) 2n = 34
ii) Aegle marmelos	b) 2n = 18
iii) Cydonia oblonga	c) 2n = 26
iv) Prunus armeniaca	d) 2n = 22
v) Psidium guajava	e) 2n = 16

136. Match the following Apple varieties and their attributes:

Apple variety	Attributes
i) Prima	a) Spur type
ii) Parlin's Beauty	b) Mutant of Delicious
iii) Red Chief	c) Indigenous variety
iv) Ambri	d) Low chilling type
v) Vance Delicious	e) Scab resistant

137. Match the following Mango varieties and their corresponding attributes:

Mango variety	Attributes
i) Pusa Shresth	a) Seedling of Haden
ii) Sindhu	b) Free from spongy tissue
iii) Arka Puneet	c) Elongated fruits with red peel
iv) Osteen	d) Thin stone with non-viable kernel
v) Neldica	e) Seedling of Palmer

138. Match the native fruit names with their corresponding scientific names:

i) Calamondin	a) *Citrus reshmi*
ii) Cleopatra mandarin	b) *Citrus pennivesiculata*
iii) Gajanimma	c) *Citrus jambhiri*
iv) Rough lemon	d) *Citrus latifolia*
v) Tahiti lime	e) *Citrus madurensis*

139. Match these resistant

Resistant Vitis	Disease species
i) *V. aestivalis*	a) Downey mildew and black rot
ii) *V. vulpine*	b) Downey and powdery mildews
iii) *V. cinerea*	c) Downey mildew
iv) *V. labrusca*	d) Downey powdery mildews & black rot
v) *V. smalliana*	e) Pierce's disease

140. Match the given fruits with their respective places of origin:

Fruit		Place of origin	
i)	Loquat	a)	Sunda Island
ii)	Pomegranate	b)	Tropical America
iii)	Fig	c)	Turkey
iv)	Mangosteen	d)	China
v)	Annonaceous fruits	e)	Iran

Answers Key

1.	(a)	2.	(c)	3.	(d)	4.	(a)	5.	(b)	6.	(c)	7.	(b)
8.	(d)	9.	(d)	10.	(a)	11.	(b)	12.	(d)	13.	(a)	14.	(c)
15.	(b)	16.	(c)	17.	(a)	18.	(b)	19.	(a)	20.	(c)	21.	(c)
22.	(b)	23.	(c)	24.	(a)	25.	(c)	26.	(d)	27.	(c)	28.	(b)
29.	(d)	30.	(a)	31.	(a)	32.	(b)	33.	(b)	34.	(a)	35.	(b)
36.	(c)	37.	(d)	38.	(c)	39.	(a)	40.	(a)	41.	(d)	42.	(a)
43.	(c)	44.	(d)	45.	(b)	46.	(b)	47.	(b)	48.	(c)	49.	(c)
50.	(a)	51.	(a)	52.	(c)	53.	(a)	54.	(a)	55.	(c)	56.	(a)
57.	(a)	58.	(a)	59.	(a)	60.	(b)	61.	(b)	62.	(d)	63.	(c)
64.	(b)	65.	(c)	66.	(a)	67.	(c)	68.	(c)	69.	(d)	70.	(b)
71.	(c)	72.	(c)	73.	(c)	74.	(b)	75.	(b)	76.	(b)	77.	(d)
78.	(c)	79.	(a)	80.	(c)	81.	(a)	82.	(d)	83.	(a)	84.	(b)
85.	(c)	86.	(b)	87.	(d)	88.	(a)	89.	(b)	90.	(b)	91.	(b)
92.	(b)	93.	(d)	94.	(c)	95.	(c)	96.	(b)	97.	(c)	98.	(b)
99.	(a)	100.	(a)	101.	(a)	102.	(d)	103.	(d)	104.	(a)	105.	(a)
106.	(d)	107.	(b)	108.	(c)	109.	(c)	110.	(c)	111.	(a)	112.	(a)
113.	(a)	114.	(b)	115.	(a)	116.	(c)	117.	(a)	118.	(d)	119.	(a)
120.	(c)	121.	(b)	122.	(b)	123.	(b)	124.	(d)	125.	(b)	126.	(b)
127.	(a)	128.	(d)	129.	(d)	130.	(b)						

Q.	131.	132.	133.	134.	135.	136.	137.	138.	139.	140.
i)	(c)	(c)	(c)	(c)	(c)	(e)	(c)	(e)	(c)	(d)
ii)	(d)	(e)	(a)	(d)	(b)	(d)	(d)	(a)	(a)	(e)
iii)	(e)	(a)	(e)	(a)	(a)	(a)	(b)	(b)	(d)	(c)
iv)	(b)	(b)	(b)	(e)	(e)	(c)	(a)	(c)	(b)	(a)
v)	(a)	(d)	(d)	(b)	(d)	(b)	(e)	(d)	(e)	(b)

4

ICAR – IARI Ph.D.
Fruit Science Entrance Exam – 2015

1. DD Kisan was launched on:
   a) 23 May 2015 b) 26 May 2015
   c) 29 May 2015 d) 16 May 2015
2. UN General Assembly declared 2015 as:
   a) International Year of Soil
   b) International Year of Pulses
   c) International Year of Sustainable Agri.
   d) International Year of Family Farming
3. As per the advance estimates for 2014-15, the total food grain production in the country is estimated at:
   a) 247 million tonnes b) 253 million tonnes
   c) 257 million tonnes d) 264 million tonnes
4. Highest number of applications were received by the PPV&FR Authority for registration and protection under the Act for:
   a) Cereals b) Vegetables
   c) Fruit crops d) Ornamentals
5. Which of the following is the richest source of vitamin C (ascorbic acid)?
   a) Lime b) Guava
   c) Aonla d) Ber
6. HD-2967 is a variety of:
   a) Barley b) Wheat
   c) Rye d) Rice
7. Which of the following oilseed crops covers maximum cultivation area under India?
   a) Groundnut b) Soybean
   c) Mustard d) Sesame

8. The gene responsible for 'Green revolution' is related with which of the following hormones?
   a) Auxin b) Cytokinin
   c) Gibberellin d) Abscisic acid
9. The traditional cleaning of rice and wheat is known as:
   a) Threshing b) Winnowing
   c) Sieving d) Drying
10. Which of the following is not a measure of central tendency?
   a) Menu b) Mode
   c) Median d) Standard deviation
11. Arithmetic mean and variance are always equal (Mean=variance) in:
   a) Normal distribution b) Poisson distribution
   c) Nominal distribution d) None of the above
12. Which of the following is a cross pollinated crop?
   a) Wheat b) Rice
   c) Sorghum d) Pearl millet
13. Which of the following is a water-soluble vitamin?
   a) Vitamin A b) Vitamin D
   c) Vitamin C d) Vitamin E
14. Water use efficiency of drip irrigation is:
   a) >35% b) >55%
   c) >75% d) >85%
15. Which of the following pesticides is not banned in India?
   a) Aldrin b) Aldicarb
   c) Heptachlor d) Chlorpyrifos
16. The increasing concentration of n substance such as a toxic chemical in the tissues of organisms at successively higher levels in a food chain is known as:
   a) Bioaccumulation b) Bio-magnification
   c) Ecotoxicology d) Persistence
17. Which of the following is a re-emerging pathogen/problem in rice?
   a) Blast (*Pyricularia oryzae*)
   b) Bacterial blight (*Xanthomonas oryzae*)
   c) Sheath blight (*Rhizoctonia solani*)
   d) Brown spot (*Helminthosporium oryzae*)

18. Pulses are deficient in which essential amino acid?
    a) Lysine b) Methionine
    c) Tryptophan d) Leucine
19. Balanced fertilizer (N:P:K) ratio for pulses is:
    a) 1:2:2 b) 3:2:1
    c) 4:2:1 d) 2:1:1
20. Which of the following is a permanent physical property of soil?
    a) Soil texture b) Soil structure
    c) Soil consistence d) Soil density
21. Which of the following nutrients constitute about 90% of the total dry matter of plants?
    a) C, H, O b) N, P, K
    c) Ca, Mg, S d) Fe, Mn, Zn
22. The actual pressure with which water enters into the cell is called:
    a) DPD b) WP
    c) OP d) TP
23. Leverage ratio is:
    a) Differed liabilities/ Net worth b) Net worth/ Differed liabilities
    c) Net worth/ Total assets d) None of these
24. Which form of sugar is absent in *Vitis vinifera* at the time of ripening?
    a) Sucrose b) Fructose
    c) Glucose d) Dextrose
25. Viviparous seeds are reported in:
    a) Mango b) Jackfruit
    c) Grapes d) Papaya
26. Which of the following is the richest source of vitamin C?
    a) Barbados cherry b) Guava
    c) Aonla d) Ber
27. Double working is commonly followed in:
    a) Plum on its own roots b) Pear on quince rootstock
    c) Mango on 13-1 rootstock d) Apple on crab apple rootstock
28. Gutta Pecha originated in:
    a) South America b) South Africa
    c) South Asia d) South Australia

29. Epigeal germination is seen in:
    a) Cashew nut b) Banana
    c) Jackfruit d) Citrus
30. Self-sterile (reflexed stamen) varieties of grapes are:
    a) Angur Kalan b) Hur
    c) Banque Abyad d) All of these
31. The term 'Apomixis' was coined by:
    a) Webber b) Winkler
    c) Wester d) Swingle
32. Polyembryony was first discovered by:
    a) Strasburger b) Leeuwenhoek
    c) Swingle d) Winkler
33. Mango variety 'Pusa Peetamber' is a cross of:
    a) Amrapali × Sensation b) Dashehari × Sensation
    c) Amrapali × Lal Sundari d) Lal Sundari × Amrapali
34. Which of the following *Mangifera* species bear fruits in off season?
    a) *Mangifera pentandra* b) *Mangifera lagenifera*
    c) *Mangifera rufocostata* d) *Mangifera similis*
35. The concept of centre of diversity was given by:
    a) De Candolle b) Harlan and De wet
    c) N.I. Vavilov d) Zeven and Zhukovasky
36. Which of the following *Mangifera* species has labyrinthine seed?
    a) *Mangifera griffithii* b) *Mangifera monandra*
    c) *Mangifera paludosa* d) *Mangifera gedebe*
37. The florigenic promoter, which determines flowering in mango is synthesized in:
    a) Roots b) Apical buds
    c) Flower buds d) Leaves
38. The genome size of mango is:
    a) 420 mbp b) 450 mbp
    c) 470 mbp d) 496 mbp

39. Polyembryony in mango is governed by:

a)	Single recessive gene	b)	Single dominant gene
c)	Additive gene	d)	Polygene

40. Albedo in citrus fruits is rich in:

a)	Protein	b)	Oil
c)	Cellulose	d)	All of the above

41. Tip layering is a natural method of propagation in which crop?

a)	Dewberry	b)	Raspberry
c)	Strawberry	d)	None of the above

42. Rind of citrus fruits becomes coarser and thicker due to deficiency of which element?

a)	Nitrogen	b)	Zinc
c)	Phosphorus	d)	Magnesium

43. The mango cultivar R2E2 developed in Australia is a seedling selection of:

a)	Kensington	b)	Kent
c)	Haden	d)	Turpentine

44. Black nose is a physiological disorder of date palm caused due to high humidity. This occurs during which of the following developmental stages?

a)	Gandora stage	b)	Doka stage
c)	Find stage	d)	Dang stage

45. Duration of Hababock (Habamboo) stage in date palm is:

a)	Pollination to 4 weeks	b)	4 to 5 weeks
c)	4 to 9 weeks	d)	1 week

46. Date palm improvement programme in India was first started at:

a)	Bikaner (Rajasthan)	b)	Jaisalmer (Rajasthan)
c)	Abohar (Punjab)	d)	Ludhiana (Punjab)

47. Deglet Noor variety of date palm originated from:

a)	Morocco	b)	East Africa
c)	Iraq	d)	Algeria

48. Klue Home Thong is a mutant variety of:

a)	Papaya	b)	Litchi
c)	Mango	d)	Banana

49. Jojo, the first plum pox virus resistant cultivar developed in Germany belongs to:

a) *Prunus salicina* b) *Prunus armeniaca*

c) *Prunus domestica* d) *Prunus insititia*

50. Which of the following lime cultivars is resistant to citrus canker?

a) ALH-77 b) Vikram

c) Phule Sharbati d) CRH-5

51. RTCP is a variety of:

a) Papaya b) Jackfruit

c) Custard apple d) Pineapple

52. Which of the following apple species is a progenitor of domesticated apple:

a) *Malus orientalis* b) *Malus sylvestris*

c) *Malus sieversii* d) *Malus prunifolia*

53. Ripening temperature of banana is:

a) 6-9 °C b) 9-13 °C

c) 14-20 °C d) 20-25 °C

54. Which of the following fruit crops is day neutral in nature?

a) Custard apple b) Persimmon

c) Citrus d) Grape

55. PSM-1 is a rootstock of:

a) Apple b) Cherry

c) Plum d) Apricot

56. Gomera series rootstocks of mango were developed in which country?

a) Florida b) Israel

c) South Africa d) Spain

57. Forner Alcaide (F&A) series rootstocks of citrus were developed in which country?

a) Florida b) Israel

c) South Africa d) Spain

58. Longan berry is developed from cross between:

a) Blackberry × Raspberry b) Blueberry × Raspberry

c) Blackberry × Cranberry d) Blueberry × Goji Berry

59. TRY (G) -1 is a variety of:
    a) Grape b) Guava
    c) Grapefruit d) Jackfruit
60. PLR (J) - 1 is a variety of:
    a) Jamun b) Jackfruit
    c) Jaboticaba d) Karonda
61. Which of the following gene transfer methods was used in transgenic papaya?
    a) Agrobacterium mediated b) Shot gun method
    c) Particle bombardment d) Micro projectile bombardment
62. St. Julien rootstock of plum is a selection from:
    a) *Prunus americana* b) *Prunus domestica*
    c) *Prunus insititia* d) *Prunus cerasifera*
63. Pixy, a dwarfing rootstock of plum is seedling selection of:
    a) *Prunus insititia* b) *Prunus cerasifera*
    c) *Prunus americana* d) *Prunus communis*
64. Which of the following species is a progenitor of grape?
    a) *Vitis vinifera* ssp. sylvestris b) *Vitis vinifera* ssp. caucasus
    c) *Vitis vinifera* ssp. typical d) *Vitis vinifera* ssp. balcanica
65. Which of the following is the smallest family?
    a) Passifloraceae b) Punicaceae
    c) Lauraceae d) Myrtaceae
66. Which of the following citrus species is almost thornless?
    a) *Citrus maxima* b) *Citrus paradisi*
    c) *Citrus latifolia* d) *Citrus medica*
67. Chicklet is a product of:
    a) Sapota b) Guava
    c) Custard apple d) Avocado
68. Which of the following *Actinidia* species is known as 'Baby Kiwi'?
    a) *Actinidia chinensis* b) *Actinidia arguta*
    c) *Actinidia eriantha* d) *Actinidia deliciosa*
69. The best growing altitude for cashew nut is:
    a) 700m MSL b) 400m MSL
    c) 1000m MSL d) 200m MSL

70. Which of the following fruit crops has the highest ascorbic content?

a) Barbados cherry b) Aonla

c) Guava d) Kiwifruit

71. Which of the following fruit crops is referred to as strictly subtropical owing to its exact climate requirement?

a) Litchi b) Olive

c) Loquat d) Persimmon

72. Viroid's induced dwarfing is done in:

a) Cherry b) Peach

c) Citrus d) Grape

73. 100 % fruit set retain till maturity can be seen in which fruit crop?

a) Mango b) Kiwifruit

c) Loquat d) Aonla

74. 'Naked seeds' are found in:

a) Angiosperms b) Gymnosperms

c) Both d) None

75. Juglone is synthesized in walnut in:

a) Leaves b) Roots

c) Stem d) Fruit

76. 'Apalachee' is a variety of:

a) Walnut b) Peacanut

c) Apricot d) Almond

77. Which of the following is a dwarfing rootstock of cherry?

a) F12-1 b) Mazzard

c) Colt d) Gisela

78. Species having highest PPO (Polyphenol oxidase) activity is highly resistant to:

a) Drought b) Frost

c) Flood d) Salinity

79. The most widely used seedling rootstock of walnut in India is:

a) *Juglans regia* b) *Juglans hindsii*

c) *Juglans californica* d) Paradox

80. Cracking of fruit peel is a maturity index of:
   a) Walnut b) Cherry
   c) Litchi d) Bael

81. Recurrent flowering is a recently reported physiological disorder of........mango.
   a) Alphonso b) Hedan
   c) Pairi d) Kesar

82. Which of the following is a termite resistant rootstock of jamun?
   a) *Syzygium cumini* b) *Syzygium densiflorum*
   c) *Syzygium uniflora* d) *Syzygium zeylanicum*

83. Which of the following is known as 'Mustang grape'?
   a) *Vitis amurensis* b) *Vitis munsoniana*
   c) *Vitis flexuosa* d) *Vitis davidii*

84. Which variety of grape forms inverted bottle neck when grafted on Dogridge rootstock?
   a) Thompson Seedless b) Beauty Seedless
   c) Perlette d) Ruby Seedless

85. Hypersensitivity is governed by:
   a) Single dominant gene b) Single recessive gene
   c) Polygenes d) Complementary gene

86. Causal organism of citrus variegated chlorosis is:
   a) Virus b) Bacteria
   c) MLO d) Nutrient deficiency

87. Late leafing character is an important criterion for improvement of:
   a) Almond b) Peach
   c) Cherry d) Apricot

88. 'Supernova' is a mutant variety of:
   a) Almond b) Apple
   c) Grape d) Banana

89. Which of the following varieties of mango has sweet and sour flavour with milky aroma?
   a) Palmer b) Tango
   c) Naomi d) Hindi

90. Cherry leaf roll is caused through:
    a) Aphid b) Pollen
    c) Both of the above d) None of these
91. Peach disease is caused by
    a) Bacteria b) Phytoplasma
    c) Fungi d) Virus
92. The inflorescence of litchi is:
    a) Determinate b) Indeterminate
    c) Semi-determinate d) None of these
93. Somatic chromosome number of cnycinu variety of pineapple is:
    a) 2n=2x=50 b) 2n=3x=75
    c) 2 n=4x= 100 d) 2n=6x= 150
94. Botanically, summer grape is known as:
    a) *Vitis rupestris* b) *Vttis longi*
    c) *Vitis champinii* d) *Vitis aestivalis*
95. Which of the following crops shows allelopathy?
    a) Peacanut b) Cherry
    c) Walnut d) Plum
96. Citrus decline is also known as:
    a) Dieback b) Neglectosis
    c) Chlorosis d) All of these
97. Cherry rootstock 'Paja' is botanically known as:
    a) *Prunus padus* b) *Prunus cerasoides*
    c) *Prunus avium* d) *Prunus cerasus*
98. Which of the following is an evergreen species of pear
    a) *Pyrus fauriei* b) *Pyrus chiftignifoliti*
    c) *Pyrus kochi* d) *Pyrus bretschneideri*
99. 'Orablanco' and 'Melogold' are the varieties of:
    a) Puminelo b) Grapefruit
    c) Pummclo × Grapefruit d) Grapefruit × Pummelo
100. Bizzare orange- a graft chimera has been evolved as:
    a) Sour orange and citron b) Sweet orange and citron
    c) Sweet orange and sour orange d) Sour orange and grape fruit

101. Match the following w.r.t. Cherry rootstock:

Rootstock		Characteristic	
i)	Paja	a)	Do not require chilling treatment
ii)	Mazzard	b)	Most vigorous
iii)	F12/1	c)	Very vigorous clonal rootstock
iv)	Colt	d)	Dwarfing clonal rootstock
v)	Mahaleb	e)	Deeper root system

102. Match the following w.r.t. the physiological disorder of mango cultivars:

Physiological disorder		Susceptible cultivar	
i)	Recurrent flowering	a)	Pico
ii)	Black flesh	b)	Tommy Atkms
iii)	Lumpy tissue	c)	Alphonv
iv)	Jelly seed	d)	Haden
v)	Riccy tissue	c)	Carabao

103. Match the following w.r.t. the physiological disorder of citrus cultivation:

Physiological disorder		Susceptible cultivar	
i)	Oleocellosis	a)	Rainbow mandarin
ii)	Superficial rind pitting	b)	Shamouti orange
iii)	Granulation	c)	Valencia orange
iv)	Puffiness	d)	Satsuma mandarin
v)	Membranous	e)	Lemon

104. Match the following diseases and their causal organisms:

Disease		Causal organisms	
i)	Aonla ring rust	a)	Elsinoe spp.
ii)	Ber-black leaf spot	b)	Phaeophleospora spp.
iii)	Grape-anthracnose	c)	Ravonclis spp.
iv)	Sapota-leaf spot	d)	Phytophthora spp.
v)	Cocoa-black pod	e)	Ixoriopsis sp

105. Match the following fruit crops with their chromosome numbers:

Fruit crop		Chromosome number	
i)	Phalsa	a)	18
ii)	Datepalm	b)	28
iii)	Aonla	c)	16
iv)	Apple	d)	34
v)	Almond	e)	36

106. Match the following varieties of fruit crops with ....:

i)	Papaya	a)	Rainbow
ii)	Banana	b)	Honey Sweet
iii)	Grape	c)	Honey Crips
iv)	Apple	d)	Chardonnay
v)	Plum	e)	Novaria

107. Match the following fruit crops with their related terms:

i)	Peach	a)	Bitter pit
i)	Almond	b)	J.H. Hale
iii)	Plum	c)	Colt
iv)	Apple	d)	Pear less
v)	Cherry	e)	President

108. Match the following fruit crop with their common names:

i)	*Citrus paradisi*	a)	Grape fruit
ii)	*Citrus medico*	b)	Pummelo
iii)	*Citrus limon*	c)	Sweet orange
iv)	*Citrus sinensis*	d)	Citron
v)	*Citrus grandis*	e)	Lemon

109. Match the following:

Family		Scientific name	
i)	Fabaceae	a)	*Physalis peruviana* L.
ii)	Combretaceae	b)	*Persea americana* M.
iii)	Solanaceae	c)	*Terminalia catappa* L.
iv)	Lauraceae	d)	*Nephelium lappaceum* L.
v)	Sapindaceae	e)	*Tamarindus indica* L.

110. Match the following:

i)	Protoandry	a)	Avocado
ii)	Protogyny	b)	Chestnut
iii)	PDSD	c)	Sapota
iv)	Dichogamy	d)	Pecan nut
v)	Heterodichogamy	e)	Walnut

Answers Key

1.	(b)	2.	(a)	3.	(b)	4.	(a)	5.	(c)	6.	(b)	7.	(b)
8.	(c)	9.	(b)	10.	(d)	11.	(b)	12.	(d)	13.	(c)	14.	(d)
15.	(d)	16.	(b)	17.	(c)	18.	(b)	19.	(a)	20.	(a)	21.	(a)
22.	(c)	23.	(a)	24.	(a)	25.	(b)	26.	(a)	27.	(b)	28.	(a)
29.	(a)	30.	(d)	31.	(b)	32.	(b)	33.	(c)	34.	(c)	35.	(c)
36.	(d)	37.	(d)	38.	(b)	39.	(b)	40.	(c)	41.	(a)	42.	(c)
43.	(b)	44.	(b)	45.	(a)	46.	(c)	47.	(d)	48.	(d)	49.	(c)
50.	(a)	51.	(a)	52.	(c)	53.	(c)	54.	(c)	55.	(a)	56.	(d)
57.	(d)	58.	(a)	59.	(b)	60.	(b)	61.	(d)	62.	(c)	63.	(a)
64.	(a)	65.	(b)	66.	(a)	67.	(a)	68.	(b)	69.	(a)	70.	(a)
71.	(c)	72.	(c)	73.	(b)	74.	(b)	75.	(a)	76.	(b)	77.	(c)
78.	(b)	79.	(a)	80.	(a)	81.	(a)	82.	(b)	83.	(b)	84.	(a)
85.	(a)	86.	(b)	87.	(a)	88.	(a)	89.	(b)	90.	(b)	91.	(b)
92.	(a)	93.	(b)	94.	(d)	95.	(c)	96.	(d)	97.	(b)	98.	(c)
99.	(c)	100.	(a)										

Q.	101.	102.	103.	104.	105.	106.	107.	108.	109.	110.
i)	(a)	(c)	(a)	(c)	(a)	(a)	(b)	(e)	(e)	(e)
ii)	(b)	(d)	(b)	(e)	(e)	(e)	(d)	(d)	(c)	(c)
iii)	(c)	(a)	(c)	(a)	(b)	(d)	(e)	(e)	(a)	(a)
iv)	(d)	(b)	(d)	(b)	(d)	(c)	(a)	(c)	(b)	(b)
v)	(e)	(c)	(e)	(d)	(c)	(b)	(c)	(b)	(d)	(d)

5

ICAR – IARI Ph.D. Fruit Science Entrance Exam – 2016

1. National Agriculture Market (NAM), an online trading portal for farm produce was launched on:
   a) 4 April 2016 b) 10 April 2016
   c) 14 April 2016 d) 16 April 2016
2. UN General Assembly declared 2016 as:
   a) International Year of Oilseeds
   b) International Year of Pulses
   c) International Year of Sustainable Agri.
   d) International Year of Cereals
3. Which of the following is the richest source of vitamin C (ascorbic acid)?
   a) Lime b) Guava
   c) Aonla d) Ber
4. The most critical stage of irrigation in wheat crop is:
   a) CRI stage b) Flowering stage
   c) Tillering stage d) Jointing stage
5. Which of the following combination of crops is C4 Plant?
   a) Wheat and rice b) Soybean and groundnut
   c) Cotton and sunflower d) Maize and sugarcane
6. Cotton belongs to the ....family:
   a) Cruciferae b) Anacardiaceae
   c) Malvaceae d) Solanaceae
7. Which one of the following is a Rabi crop?
   a) Pearl millet b) Soybean
   c) Maize d) Chick pea

8. Khaira disease of rice is due to the deficiency of:
   a) Fe b) Mg
   c) B d) Zn
9. For applying 25 kg of nitrogen, how much urea would one use?
   a) 45 kg b) 50 kg
   c) 55 kg d) 60 kg
10. The share of agriculture and allied activities in India's GDP at constant prices in 2013-14 was:
   a) 14% b) 15%
   c) 16% d) 17%
11. India's rank in world vegetable production is:
   a) First b) Second
   c) Third d) Fourth
12. Which of the following diseases affects the export of wheat from India?
   a) Black rust b) Loose smut
   c) Hill bunt d) Karnal bunt
13. Red colour of tomato is due to:
   a) Anthocyanin b) Lycopene
   c) Carotene d) Prolycopene
14. Which of the following is a measure of dispersion?
   a) Mean b) Median
   c) Mode d) Range
15. The largest and the most important soil group of India is:
   a) Alluvial soils b) Black Soils
   c) Lateritic Soils d) Desert Soils
16. ARYA stands for:
   a) Attracting and Retaining Youth in Agriculture
   b) Attaining and Retraining Youth in Agriculture
   c) Attending and Retraining Youth in Agriculture
   d) Adopting and Retraining Youth in Agriculture
17. The chairman of the governing board of ATMA (Agriculture Technology Management Agency) is:
   a) Gram Sevak b) Tehsildar
   c) BDO d) District Magistrate

18. *Caenorhabditis elegans* is a

    a) Weed b) Fungi

    c) Nematode d) Bacteria

19. Maize protein is known as:

    a) Gluten b) Zein

    c) Casein d) Globulin

20. Which of the following plant hormones plays an important role in seed germination?

    a) Auxin b) Cytokinin

    c) Gibberellin d) ABA

21. Rhizobium species responsible for nitrogen fixation in chickpea is:

    a) *Bradyrhizobium japonicum* b) *Rhizobium meliloti*

    c) *Rhizobium leguminosarum* d) *Rhizobium ciceri*

22. White flies are the principle vectors of:

    a) Leaf curl viruses b) Spotted wilt viruses

    c) Ring spot viruses d) Fan leaf viruses

23. Which of the following is a primary tillage implement?

    a) Mould-board plough b) Harrow

    c) Leveller d) Ridger

24. is the source of dwarfing gene in wheat.

    a) Dee-Gee-Woo-Gen b) Norin-10

    c) Combine Kofir-60 d) Opaque-2

25. At harvest, the above ground straw (leaf, sheath and stem) weight and grain weight of paddy crop are 6.0 and 4.0 tonnes per hectare, respectively. What is the harvest index of paddy?

    a) 40% b) 50%

    c) 60% d) 90%

26. Virus free plants can be obtained through:

    a) Anther culture b) Meristem culture

    c) Embryo culture d) Ovule culture

27. November dump is a physiological disorder banana occurs due to

    a) Low temperature b) High temperature

    c) Water logging d) Toxicity of Cl ions

28. Flavouring compound in jamun is

a) Methyl propionate
b) Methyl salicylate
c) Methyl anthranilate
d) Methyl butanoate

29. TRY (G) -1 is a variety of:

a) Grape
b) Guava
c) Grapefruit
d) Jackfruit

30. Which of the following apple varieties is not a true delicious

a) Golden Delicious
b) Royal Delicious
c) Mollies Delicious
d) Starking Delicious

31. In recent classification of mango which of the following sections of Mangifera has maximum number of species?

a) *Marchandara* Pierre
b) *Euantherae* Pierre
c) *Rawa* Kosterm
d) *Mangifera* Ding Hou

32. Which of the following Mangifera species give free stone mango?

a) *Mangifera magnifica*
b) *Mangifera pajang*
c) *Mangifera similis*
d) *Mangifera altissima*

33. Mangifera species in which flowering occurs once in every 5-10 years?

a) *Mangifera similis*
b) *Mangifera casturi*
c) *Mangifera foetida*
d) *Mangifera lagenifera*

34. International Biodiversity Day is celebrated on:

a) 23 November
b) 22 May
c) 22 December
d) 23 April

35. Which of the following date palm species is most dwarfing:

a) *Phoenix reclinata*
b) *Phoenix acqualis*
c) *Phoenix humilis*
d) *Phoenix sylvestris*

36. Which of the following honey bee species is considered most efficient and abundant pollinator?

a) *Apis cerana*
b) *Apis mellifera*
c) *Apis florea*
d) *Apis dorsata*

37. The florigenic promoter, which determines flowering in mango, is synthesized in:

a) Roots
b) Apical buds
c) Flower buds
d) Leaves

38. The genome size of mango is:

a)	420 mbp	b)	450 mbp
c)	470 mbp	d)	496 mbp

39. Papaya species resistant to *Phytophthora* is:

a)	*Vasconcellea guardian*	b)	*Vasconcellea stipulata*
c)	*Vasconcellea heliborne*	d)	*Vasconcellea pubescens*

40. Albedo petaca disorder in citrus occurs due to the deficiency of:

a)	Zinc (Zn)	b)	Boron
c)	Molybdenum (Mo)	d)	Iron (Fe)

41. PGR, New Delhi has established gene sanctuary for citrus in garo hills, Meghalaya during the year:

a)	1976	b)	1978
c)	1980	d)	1982

42. Etrog citron is an indicator of:

a)	*Citrus xyloporosis*	b)	*Citrus exocortis*
c)	*Citrus variegated* chlorosis	d)	*Citrus psorosis*

43. The mango variety 'Alpha' developed in Brazil is a hybrid of:

a)	Mallika × Van dyke	b)	Amrapali × Winters
c)	Mulgoa ×Turpentine	d)	Turpentine × Eldon

44. Sister seedling of Ratna mango is:

a)	Konkan Raja	b)	Konkan Ruchi
c)	Sonpari	d)	Sabori

45. Medica a hybrid variety of grape developed at NRC, Grape, Pune (MH). The parents of the hybrids are:

a) Pusa Navrang × Manjri Naveen

b) Arkavati × Flame Seedless

c) Pusa Navrang × Flame Seedless

d) Pusa Urvashi × Beauty Seedless

46. Balaji is a variety of:

a)	Lemon	b)	Acid lime
c)	Pummelo	d)	Grapefruit

47. Spanish lime (*Melicoccus bijugatus*) belongs to ....family:

a)	Sapindaceae	b)	Rutaceae
c)	Ebenaceae	d)	Punicaceae

48. Daru a wild type pomegranate found in western Himalayan region is resistant to:
    a) Fruit cracking b) Bacterial blight
    c) Internal breakdown d) Leaf spot
49. Transgenic variety of plum resistant to plum pox virus:
    a) Honey Crips b) Honey Sweet
    c) Jojo d) President
50. Cold hardy species of *Psidium* is:
    a) *Psidium molle* b) *Psidium guinenese*
    c) *Psidium montanum* d) *Psidium cattleianum*
51. Arka Prabhat a papaya hybrid is a result of cross between:
    a) Sunrise Solo × Pink Flesh Sweet
    b) Surya × Taichung × Local Dwarf
    c) Sunrise Solo × Local Dwarf
    d) Sunrise Solo × Subhang 6
52. Which of the following Musa species is most cold hardy:
    a) *Musa basjoo* b) *Musa balbisiana*
    c) *Musa acuminata* d) *Musa velutina*
53. Ripening temperature of banana is:
    a) 6-9 °C b) 9-13 °C
    c) 14-20 °C d) 20-25 °C
54. Which of the following pear species is known as snow pear?
    a) *Pyrus nivalis* b) *Pyrus caucasica*
    c) *Pyrus calleryana* d) *Pyrus salicifolia*
55. Black apricot is botanically known as:
    a) *Prunus britannica* b) *Prunus ussuriensis*
    c) *Prunus dasycarpa* d) *Prunus mandshurica*
56. INIBAP (now Biodiversity International Banana and Plantain Section) was established in:
    a) 1975 b) 1980
    c) 1985 d) 1995
57. November dump, a physiological disorder of banana occurs due to:
    a) Low temperature b) High temperature
    c) Water logging d) Toxicity of Cl ions

58. Flavouring compound in jamun is:

a) Methyl propionate b) Methyl salicylate

c) Methyl anthranilate d) Methyl butanoate

59. TRY (G) -1 is a variety of:

a) Grape b) Guava

d) Grapefruit d) Jackfruit

60. Which of the following apple varieties is not a true delicious:

a) Golden Delicious b) Royal Delicious

c) Mollies Delicious d) Starking Delicious

61. Protogynous diurnally synchronous geitonogamous flowering pattern occurs in:

a) *Annona muricata* b) *Castanea sativa*

c) *Pistacia vera* d) *Persea americana*

62. Dichotomous branching habit is found in:

a) Aonla b) Lasoda

c) Jackfruit d) Karonda

63. Which of the following avocado race has/ have maximum oil content:

a) Mexican race b) Guatemalan race

c) West Indian race d) All the above

64. International Horticulture Congress-2018 of ISHS was held at:

a) Lisbon b) Paris

c) Turkey d) Brisbane

65. *Persea schiedeana* is resistant to:

a) Heart rot b) Anthracnose

c) Black spot d) Cercospora spot

66. Degrain is a physiological disorder of:

a) Papaya b) Grape

c) Avocado d) Banana

67. The concept of effective pollination period (EPP) was developed by:

a) Janick b) Williams

c) Frost d) Swingle

68. Zygodormancy of aonla (*Emblica officinalis*) is due to:
    a) Ethylene b) Auxin
    c) ABA d) Gibberellin
69. Which of the following is known as 'self-peeling banana'?
    a) *Musa laterita* b) *Musa ingens*
    c) *Musa basjoo* d) *Musa velutina*
70. 'Suppressed climacteric' varieties are found in:
    a) Apple b) Banana
    c) Plum d) Peach
71. Red colour in onion is due to ....pigment:
    a) Carotene b) Quercetin
    c) Lycopene d) Anthocyanin
72. In tamarind, folding of the leaflets after dark is attributed to:
    a) Free tartaric acid b) Free fatty acid
    c) Lupeol d) ADF bound nitrogen
73. Ambri is not grown in other parts of country except Kashmir because of its:
    a) High chilling requirement b) Late maturity
    c) Extra ordinary keeping quality d) Indigenous to Kashmir
74. Which of the following is not a valid species of citrus?
    a) *Citrus reticulata* b) *Citrus maxima*
    c) *Citrus medico* d) *Citrus latipes*
75. In which fruit crop, is the TECF precocity gene found?
    a) Peach b) Strawberry
    c) Cranberry d) Plum
76. Spiny leaves in pineapple are:
    a) Recessive b) Dominant
    c) Polygenic d) All the above
77. Cocopeat has per cent porosity.
    a) 15-20% b) 20-25%
    c) 25-30% d) 30-35%
78. Wetting period required for apple scab infection is:
    a) 17.2-23.9°C for 9 hrs b) 12.5-15.5°C for 9 hrs
    c) 24.7-29.6°C for 9 hrs d) 30.7-35.9°C for 9 hrs

79. Which of the following is not a good chilling model?
    a) Winter chilling model
    b) Utah chilling model
    c) Dynamic chilling model
    d) Shimla chilling model
80. Ink disease is one of the most destructive diseases of:
    a) Chestnut
    b) Hazelnut
    c) Peacanut
    d) Walnut
81. Indicator plant for rubbery wood disease is:
    a) Apple
    b) Cherry
    c) Walnut
    d) Almond
82. Subtropical pear cultivar having brown flesh is:
    a) Golden Nijjisekki
    b) Punjab Gold
    c) Punjab Nectar
    d) Osa Gold
83. Mattocking practice of banana has originated in:
    a) Brazil
    b) Malaysia
    c) Honduras
    d) Jamaica
84. The banana variety having horizontal bunch orientation is:
    a) Nendran
    b) Rasthali
    c) Grand Naine
    d) Monthan
85. GA4 and GA7 in apple seeds causes inhibitory effects known as:
    a) Ecodormancy
    b) Endodormancy
    c) Paradormancy
    d) Embryonic dormancy
86 Pericarp browning of litchi (*Litchi chinensis*) is due to degradation of:
    a) Anthocyanin
    b) Carotene
    c) Xanthophyll
    d) Lycopene
87. Which of the following grape varieties is used for 'champagne' making?
    a) White Riesling
    b) Pinot Noir
    c) Chardonnay
    d) Tempranillo
88. Colouring compound in cherry is:
    a) Methyl salicylate
    b) Keracyanin
    c) Cyanidin
    d) None of the above
89. Mango breeding programme was first started in the world in:
    a) India
    b) Florida
    c) Brazil
    d) Israel

90. Raki is a product of:
    a) Grape b) Fig
    c) Date palm d) Cherry
91. Christmas tree appearance of pear is due to the deficiency of:
    a) Ca b) Fe
    c) Mg d) N
92. Apple disorder in gladiolus is because of deficiency:
    a) Boron b) Calcium
    c) Nitrogen d) Potassium
93. Negative geotropism, a disorder in gladiolus occurs during:
    a) Harvesting b) Packing
    c) Transportation d) Selling
94. Indicator plant for SO2 pollution is:
    a) Gladiolus b) Ferns
    c) Dahlia d) Lichen
95. A variety suitable for greenhouse cultivation is:
    a) First Red b) Mohini
    c) Gladiator d) La France
96. The choi is a vegetable:
    a) Leafy b) Flower
    c) Stem d) Root
97. Which of the following is not a measure of central tendency:
    a) Arithmetic mean b) Median
    c) Mode d) Range
98. The relation between AM, GM, and HM is:
    a) AM<HM>GM b) AM>HM>GM
    c) AM<HM<GM d) AM>HM<GM
99. The allocation of the treatments to the different experimental units by a random process is known as:
    a) Replication b) Randomization
    c) Local control d) None of the above

100. The office of 'Geographical Indications' is located at:

a) New Delhi b) Mumbai

c) Chennai d) Kolkata

101. Pineapple was first introduced in India at:

a) Assam b) West Bengal

c) Goa d) Kerala

102. Mango was first introduced in which country?

a) Brazil b) West Indies

c) Cuba d) Florida

103. Avocado was introduced in India in :

a) 1935 b) 1942

c) 1974 d) 1984

104. Madhu Angoor is a selection from:

a) *Vitis vinifera* b) *Vitis labrusca*

c) *Vitis parviflora* d) *Vitis lanata*

105. Critical day/night temperature for flowering in mango is:

a) 30/20°C b) 25/20°C

c) 20/15°C d) 15/10°C

106. Berry size of Sharad Seedless grape for export is:

a) 12-15 mm b) 15-18 mm

c) 18-21 mm d) 21-25 mm

107. The papain having the activity of..........tyrosine units is most preferred....... in world trade.

a) 400-600 b) 600-800

c) 800-1000 d) 1000-1200

108. The largest pineapple germplasm collection made worldwide is at:

a) MARDI-Malaysia b) CIRAD-France

c) INIBAP-France d) PRI-Hawaii

109. Long term storage of lemon at low temperature (<13°C) causes chilling injury, a physiological disorder in which internal browning develops in the membrane of the segments and core is known as:

a) Oleocellosis b) Peteca

c) Creasing d) Membranosus

110. Which of the following genera has sessile flower?

a) *Citrus* b) *Fortunella*

c) *Poncirus* d) *Eremocitrus*

111. Black heart in pineapple appears when exposed to temperature:

a) 10-15°C b) 15-20°C

c) 20-25°C d) 25-30°C

112. Which was developed first?

a) Plumcots b) Pluots

c) Apriums d) Apriplums

113. The name 'plumcot' was created by:

a) Luther Burbank b) Floyd Zaiger

c) T.A. Knight d) A.P. De Candolle

114. Apoplexy, a physiological disorder of apricots occurs due to:

a) High temperature b) Low temperature

c) Water logging d) Drought

115. Water quality of heat tolerance is measured by:

a) Specific heat b) Specific density

c) Specific gravity d) All the above

116. ABA is synthesized in plant cell in:

a) Plastid b) Mitochondria

c) Endoplasmic reticulum d) Cytosol

117. Toxic substances found in litchi seed are:

a) Methylene cyclopropyl glycine b) Saponin

c) Methyl butanoate d) Ethyl hexanoate

118. Which of the following is/ are bio control agents:

a) *Pseudomonas fluorescens* b) *Trichoderma* spp.

c) *Cryptolaemus montrouzieri* d) All the above

119. During Vapour heat treatment of mango, number temperature is maintained at 50- 52°C while the pulp temperature should be:

a) 43°C b) 45°C

c) 47°C d) 49°C

120. Which of the following *Ziziphus* is a host for lac insect?
    a) *Ziziphus horsfieldii* b) *Ziziphus oenoplia*
    c) *Ziziphus xylopyrus* d) *Ziziphus incurva*

121. Which of the following Annona species has mango like flavour?
    a) *Annona cherimola* b) *Annona muricata*
    c) *Annona atemoya* d) *Annona diversifolia*

122. Lady is popular variety of apple. It was developed in which country?
    a) Australia b) Germany
    c) Japan d) Florida

123. Arka Rashmi is a recently released variety of guava. The parentage of this hybrid are
    a) Allahabad Safeda × Seedless b) Allahabad Safeda × Arka Mridula
    c) Purple Local × Kamsari d) Kamsari × Purple Local

124. In Navel orange, navel is present at the:
    a) Thalmic end b) Micropylar end
    c) Styler end d) None of the above

125. Enology is the study of:
    a) Monument b) Wines
    c) Drying of horticultural produce d) *In-vitro* propagation

126. Photosynthetically, active radiation (PAR) wave length range is:
    a) 0.1 to 0.3 μm b) 0.4 to 0.7 μm
    c) 0.7 to 1.0 μm d) 1.0 to 1.3 μm

127. Fruits of which plant contains some substances which has the ability to taste bitter substances sweet:
    a) *Gymnema sylvestre* b) *Synsepalum dulcificum*
    c) *Laminaria, japonica* d) *Glyrizzia indica*

128. A physiological disorder called 'Dry Neck' is common in:
    a) Sapota b) Mangosteen
    c) Tree tomato d) Avocado

129. Genome constitution of 'Nendran' banana is:
    a) AA b) AAA
    c) AAB d) ABB

130. Which one is not related to apple disease forewarning system?

a) VENTEM b) PODEM

c) BLITECAST d) NECTEM

131. Match the fruit crops with physiological disorders

Fruit crop	Physiological disorders
i) Mango	a) Bitter pit
ii) Sapota	b) Coulour
iii) Grape	c) Fascination
iv) Apple	d) Cocks comb
v) Pineapple	e) Jelly seed

132. Match the micronutrients with their function.

Micronutrient	Function
i) Boron	a) Chlorosis on younger leaves
ii) Calcium	b) Interveinal chlorosis
iii) Magnesium	c) Pollen germination
iv) Sulphur	d) Plastocyanin synthesis
v) Copper	e) Tip burning

133. Match the scientists and their contributions:

Scientist	Contribution
i) Kostermans and Bompard	a) Banana
ii) J. E. Higgins	b) Papaya
iii) W. T. Swingle	c) Citrus
iv) H. P. Olmo	d) Mango
v) Simmonds and Shephard	e) Grape

134. Match the following mango varieties and method of development.

Variety	Method
i) Eldon	a) Mutation
ii) Samar Behist Chausa	b) Clonal selection
iii) Dashehari 51	c) Introduction
iv) Mallika	d) Hybridization
v) Rosica	e) Chance seedling

135. Match the botanical names with the common names of citrus fruits:

Botanical name	Common name
i) *Citrus medica*	a) Satkara
ii) *Citrus ichangensis*	b) Icheng narang
iii) *Citrus rugulosa*	c) Attani
iv) *Citrus indica*	d) Indian wild orange
v) *Citrus macroptera*	e) Citron

136. Match the following fruit crops and problems.

Problem	Fruit crops
i) Sporophytic SI	a) Mangosteen
ii) Gametophytic SI	b) Avocado
iii) Pollen sterility	c) Loquat
iv) Parthenogenesis	d) Mango
v) Dichogamy	e) Ber

137. Match the fruits with their flavouring compounds of fruit crops.

Fruits	Flavouring compound
i) Apple	a) Ethyl-hexanoate
ii) Strawberry	b) Methyl-butyrate
iii) Cherry	c) Gerniol
iv) Raspberry	d) Methyl salicylate
v) Guava	e) a-caryophyllene

138. Match the following diseases and their causal organisms:

Disease	Causal organisms
i) Aonla ring rust	a) *Elsinoe* spp.
ii) Ber-black leaf spot	b) *Phaeophleospora* spp.
iii) Grape-anthracnose	c) *Ravonclis* spp.
iv) Sapota-leaf spot	d) *Phytophthora* spp.
v) Cocoa-black pod	e) *Ixoriopsis* sp

139. Match the following fruits with their places of origin:

Fruit crops	Place of origin
i) Litchi	a) Turkey
ii) Durian	b) Borneo
iii) Kambutan	c) Malaya archipelago
iv) Dragon fruit	d) South America
v) Fig	e) Southern China

140. Match the following:

Rootstock		Resistant against	
i)	Robusta No. 5	a)	Fire blight in apple
ii)	Northern spy	b)	Fire blight in pear
iii)	P 1,2,3	c)	Collar rot
iv)	OH x F	d)	Wolly apple aphid
v)	MM 106	e)	Cold tolerant

Answers Key

1.	(c)	2.	(b)	3.	(c)	4.	(a)	5.	(d)	6.	(c)	7.	(d)
8.	(d)	9.	(c)	10.	(a)	11.	(b)	12.	(d)	13.	(b)	14.	(d)
15.	(a)	16.	(a)	17.	(d)	18.	(c)	19.	(b)	20.	(c)	21.	(d)
22.	(a)	23.	(a)	24.	(b)	25.	(a)	26.	(b)	27.	(a)	28.	(c)
29.	(b)	30.	(a)	31.	(d)	32.	(c)	33.	(d)	34.	(b)	35.	(c)
36.	(a)	37.	(d)	38.	(b)	39.	(a)	40.	(b)	41.	(b)	42.	(b)
43.	(a)	44.	(b)	45.	(c)	46.	(b)	47.	(a)	48.	(b)	49.	(b)
50.	(d)	51.	(b)	52.	(d)	53.	(c)	54.	(a)	55.	(c)	56.	(c)
57.	(a)	58.	(c)	59.	(b)	60.	(a)	61.	(d)	62.	(d)	63.	(a)
64.	(c)	65.	(a)	66.	(d)	67.	(b)	68.	(b)	69.	(d)	70.	(c)
71.	(d)	72.	(c)	73.	(a)	74.	(d)	75.	(a)	76.	(a)	77.	(c)
78.	(a)	79.	(d)	80.	(a)	81.	(a)	82.	(a)	83.	(d)	84.	(a)
85.	(d)	86.	(a)	87.	(a)	88.	(b)	89.	(a)	90.	(a)	91.	(b)
92.	(b)	93.	(c)	94.	(d)	95.	(a)	96.	(a)	97.	(d)	98.	(b)
99.	(b)	100.	(c)	101.	(c)	102.	(a)	103.	(b)	104.	(a)	105.	(c)
106.	(b)	107.	(c)	108.	(a)	109.	(d)	110.	(c)	111.	(b)	112.	(a)
113.	(a)	114.	(b)	115.	(a)	116.	(d)	117.	(a)	118.	(d)	119.	(c)
120.	(c)	121.	(b)	122.	(a)	123.	(d)	124.	(c)	125.	(b)	126.	(b)
127.	(b)	128.	(d)	129.	(c)	130.	(c)						

Q.	131.	132.	133.	134.	135.	136.	137.	138.	139.	140.
i)	(e)	(c)	(d)	(c)	(e)	(d)	(b)	(c)	(e)	(a)
ii)	(d)	(e)	(b)	(e)	(b)	(c)	(a)	(e)	(b)	(d)
iii)	(b)	(b)	(c)	(b)	(c)	(e)	(d)	(a)	(c)	(e)
iv)	(a)	(a)	(e)	(d)	(d)	(a)	(c)	(b)	(d)	(b)
v)	(c)	(d)	(a)	(a)	(a)	(b)	(e)	(d)	(a)	(c)

6

ICAR – NET / ARS Fruit Science Exam – 2011

1. The term 'Chimera' was coined by:
   a) Webber b) Winkler
   c) Wester d) Swingle
2. What is the optimum pH for cultivation of Kinnow mandarin?
   a) 4-5 b) 5-6
   c) 6-7 d) 7-8
3. Amygdalin is found in:
   a) Apple b) Apricot
   c) Aonla d) Almond
4. Lenticels are prominent in:
   a) Apple b) Pear
   c) Cherry d) Mango
5. Ber is resistant to drought stress due to the presence of:
   a) Extensive tap root system b) Summer dormancy
   c) Pubescent leaves d) All of these
6. Based on climacteric requirement, rambutan is a classified as:
   a) Tropical b) Subtropical
   c) Temperate d) Arid
7. Flowering in aonla occurs on which type of shoots?
   a) Determinate b) Indeterminate
   c) Apical d) All of the above
8. Which of the following fruit crop has the longest pre-bearing cycle?
   a) Avocado b) Custard apple
   c) Jamun d) Mangosteen

9. Hexagonal system of planting can accommodate …….. % more plants than square system :
   a) 10 b) 15
   c) 20 d) 25
10. Wind break is effective up to a distance of ……. times the height of the plant:
   a) 2 b) 3
   c) 4 d) 5
11. Colchicine is generally used for:
   a) Overcoming self-incompatibility
   b) Chromosome doubling
   c) Mutation breeding
   d) All of these
12. Rasthali banana belongs to which genomic group?
   a) AAA b) AAB
   c) ABB d) AB
13. When the entire outer or inner layer is mutated, the chimera is known as:
   a) Periclinal chimera b) Mericlinal chimera
   c) Sectorial chimera d) Solid mutant
14. Micro-propagation technique is most successful in propagation of:
   a) Guava b) Banana
   c) Aonla d) Mango
15. Amritha is an improved variety of:
   a) Pomegranate b) Pineapple
   c) Apple d) Aonla
16. DSH-1 and DSH-2 are the hybrids of:
   a) Sapota b) Annona
   c) Tamarind d) Mango
17. Gamboge is a physiological disorder occurring in :
   a) Citrus b) Grape
   c) Litchi d) Mangosteen
18. Loquat (*Eriobotrya japonica*) is a member of ………. family:
   a) Liliaceae b) Rosaceae
   c) Apocynaceae d) Vitaceae

19. Thompson Seedless grape originated as a result of:
    a) Mutation b) Hybridization
    c) Polyploidy d) Selection
20. Guava is a native of:
    a) Mexico b) Brazil
    c) Peru d) Florida
21. Mandarin festival is celebrated in:
    a) Japan b) China
    c) Washington d) Germany
22. Pineapple takes how many months to harvest after planting?
    a) 9-12 b) 12-15
    c) 15-18 d) 18-20
23. Which of the following fruit crops is known as 'Hairy litchi'?
    a) Longan b) Rambutan
    c) Pulsan d) Durian
24. Which country is the largest producer of mangoes in the world?
    a) Vietnam b) Florida
    c) India d) Brazil
25. Fruit cracking is a major problem in fruits having .... as an edible part:
    a) Thalamus b) Aril
    c) Mesocarp d) Cotyledons
26. Which type of parthenocarpy is present in most of the grape varieties?
    a) Vegetative b) Simulative
    c) Stenospermocarpy d) All of these
27. The PGR used as a substitute of chilling requirement in many temperate fruits is:
    a) Ethrel b) kinetin
    c) NAA d) GA3
28. The predominant organic acid present in apple is:
    a) Citric acid b) Tartaric acid
    c) Oxalic acid d) Malic acid

29. Vivipary (germination of seeds in fruits while it is still attached to mother plant) is observed in:
    a) Jackfruit b) Carambola
    c) Grape d) Sapota
30. Internal break down is a physiological disorder of:
    a) Pear b) Peach
    c) Plum d) Pomegranate
31. Which of the following is a terminal bearing fruit crop?
    a) Mango b) Citrus
    c) Guava d) Peach
32. Which of the following organizations is engaged in exporting of processed products?
    a) NAFED b) APEDA
    c) NABARD d) FCI
33. Removal of terminal portion of shoots, branches or limbs leaving its basal portion is known as:
    a) Thinning out b) Heading back
    c) Skirting d) Dehorning
34. Tapka is the maturity index of:
    a) Apple b) Citrus
    c) Date palm d) Mango
35. Mulgoa gives rise to the origin of:
    a) Haden b) Irwin
    c) Eldon d) Sensation
36. Kozlowski (1971) classified fruits on the basis of:
    a) Fruit bearing habit b) Flowering habit
    c) Mode of pollination d) Photoperiodic response
37. The headquarter of International Seed Testing Association (ISTA) is located at:
    a) Geneva b) Zurich
    c) Washington d) Paris
38. Banana is a heavy feeder of:
    a) N b) P
    c) K d) S

39. Water use efficiency is highest in:
    a) C_3 plants b) C_4 plants
    c) CAM plants d) All of these
40. Solen system of training is followed in cultivar of apple.
    a) Granny Smith b) Red Delicious
    c) Fuzi d) Gala
41. Pineapple is a/an:
    a) Annual herb b) Perennial herb
    c) Annual shrub d) Perennial shrub
42. The most stable sex form in papaya is:
    a) Male b) Female
    c) Hermaphrodite d) All of these
43. Leaves show drying scorching effect due to the deficiency of:
    a) N b) P
    c) K d) Ca
44. The one-year old shoot of grape in which the fruit bunches appear is known as:
    a) Arms b) Shoots
    c) Canes d) Spurs
45. The ratio of edible to non-edible parts in mango should be:
    a) 1 : 2 b) 2 : 3
    c) 3 : 4 d) 4 : 5
46. Which variety of grape shows simulative parthenocarpy?
    a) Thompson Seedless b) Perlette
    c) Anab-e-Shahi d) Black Corinth
47. Degreening is not done in:
    a) Mango b) Banana
    c) Citrus d) Grape
48. Which of the following is an ideal intercrop in banana orchard?
    a) Papaya b) Pomegranate
    c) Guava d) Ginger
49. Florina, a scab resistant variety of apple was introduced in India from:
    a) Florida b) France
    c) Germany d) England

50. The crop which does not belong to the same family is:
    a) Mango b) Cashew
    c) Pistachio nut d) Strawberry
51. The most commonly used training system for most of the fruit crops is:
    a) Central leader b) Open center
    c) Modified central leader d) None of these
52. Flame grapefruit is a mutant of:
    a) Foster b) Thompson
    c) Hudson d) Henderson
53. Myrobalan B is a rootstock of:
    a) Apple b) Pear
    c) Cherry d) Plum
54. The compound present in grapefruit which is helpful for people suffering from diabetes is:
    a) Limonin b) Naringin
    c) Nootkatone d) Lycopene
55. Bio-regulator used as an alternative for chilling requirement is:
    a) NAA b) BA
    c) GA d) ABA
56. Albinism is a physiological disorder of:
    a) Almond b) Apricot
    c) Kiwifruit d) Strawberry
57. Alternate hearing is not so common in:
    a) Apple b) Mango
    c) Tamarind d) Citrus
58. In papaya, female plants can be produced in bulk through:
    a) Pollen culture b) Ovule culture
    c) Meristem culture d) Cell culture
59. Haploid plants are produced through:
    a) Pollen culture b) Ovule culture
    c) Meristem culture d) Cell culture
60. In India, tissue culture has been commercially exploited in:
    a) Strawberry b) Pineapple
    c) Banana d) Papaya

61. In Rosaceae family, sugars are mostly transported in the form of:
    a) Sucrose b) Sorbitol
    c) Glucose d) Fructose
62. The number of days required from full bloom to maturity for Dwarf Cavendish banana is:
    a) 90 + 5 b) 100 + 5
    c) 125 + 5 d) 130 + 5
63. EMLA Series of apple rootstocks are resistant to:
    a) Woolly apple aphid b) Fire blight
    c) Scab d) Virus
64. Most widely used apple rootstock across the world is:
    a) M-9 b) M-27
    c) MM-104 d) Northern Spy
65. Percent fruit set in custard apple is:
    a) 0.5-5% b) 5-10%
    c) 10-15% d) 15-20%
66. The hybrid formed by fusion of two somatic cells is known as:
    a) Somatic hybrid b) Cybrid
    c) Vybrid d) Chimera
67. 'Pinching off' of terminal bud is practiced in which mango cultivar?
    a) Dashehari b) Amrapali
    c) Alphonso d) Konkan Ruchi
68. The AEZ's for mango were not identified in the ...... area of Maharashtra:
    a) Pune b) Thane
    c) Ratnagiri d) Sindhudurg
69. Total number of AEZ's (Agri Export Zones) in India is:
    a) 45 b) 50
    c) 55 d) 60
70. Triple sigmoidal growth curve is observed in:
    a) Apple b) Litchi
    c) Kiwifruit d) Seedless banana
71. Which of the following is a tetraploid (4x) species of kumquat?
    a) *Fortunella margarita* b) *Fortunella japonica*
    c) *Fortunella crassifolia* d) *Fortunella hindsii*

72. The grape cultivar whose genome has been completely sequenced is:
    a) White Riesling b) Pinot Noir
    c) Cardinal d) Himrod
73. The symptoms of Ca and B deficiency are first observed on:
    a) Terminal bud b) Younger leaves
    c) Older leaves d) All of these
74. Which of the following crop is commercially grown in protected structures in India?
    a) Banana b) Papaya
    c) Pineapple d) Strawberry
75. The nutrient element which is immobile in soil, hence needs to be applied only as a basal dose is:
    a) N b) P
    c) K d) Ca
76. The first transgenic fruit crop is:
    a) Walnut b) Pear
    c) Papaya d) Plum
77. Pre-harvest fruit drop in citrus can be controlled by spray of:
    a) BA b) GA
    c) ABA d) 2,4,5-T
78. Cauliflower's flowering and fruiting habit is observed in:
    a) Carambola b) Jackfruit
    c) Cocoa d) All of these
79. In India, pineapple is mostly packed in:
    a) Corrugated fibre board b) Wooden crates
    c) Polythene lined Gunny bag d) Polythene lined bamboo baskets
80. The fruit crop which doesn't have any developed rootstock till date is:
    a) Pear b) Plum
    c) Cherry d) Litchi
81. The fruit crop which has greater biodiversity in India is:
    a) Olive b) Kiwifruit
    c) Peach d) Apricot

82. Pollination of fig flower by the *Blastophaga* wasp takes place through:
   a) Fascination b) Parthenogenesis
   c) Caprification d) Apomixis

83. Summer pruning is practiced in:
   a) Apple b) Cherry
   c) Plum d) Citrus

84. Wood apple (*Feronia limonia*) is commercially propagated by:
   a) Seed b) Cutting
   c) Budding d) Layering

85. In peach, the flowers are produced on:
   a) Current season growth b) Last year growth
   c) Spur d) Canes

86. Improvement in papaya is done mainly through:
   a) Inbreeding and sib mating b) Polyploidy
   c) Hybridization d) Mutation

87. Cardinal grape is a cross between:
   a) Tokey × Ribier
   b) Ontario × Sultana
   c) Queen of Vineyard × Black Kishmish
   d) Himrod × Ribier

88. High Gate is a mutant of which commercial variety of banana?
   a) Gross Michel b) Dwarf Cavendish
   c) Grand Naine d) Poovan

89. European plum (*Prunus domestica*) has been originated as a natural hybrid between:
   a) *Prunus cerasifera* × *Prunus spinosa*
   b) *Prunus spinosa* × *Prunus davidiana*
   c) *Prunus cerasifera* × *Prunus persica*
   d) *Prunus persica* × *Prunus davidiana*

90. M.27 rootstock was evolved from the cross between:
   a) M.13 × M.7 b) M.13 × M.9
   c) M.7 × M.13 d) M.9 × M.13

91. Seeds are set without fertilization in:
    a) Mango b) Avocado
    c) Loquat d) Mangosteen
92. G-137 an improved variety of pomegranate is a selection from:
    a) Muscat b) Ganesh
    c) Alandi d) Jyoti
93. Flemish Beauty variety of pear is:
    a) Self-sterile b) Self-fertile
    c) Cross sterile d) Cross fertile
94. Which of the following species of papaya is resistant to virus?
    a) *Vasconcellea parviflora* b) *Vasconcellea microcarpa*
    c) *Vasconcellea cauliflora* d) *Vasconcellea pentagona*
95. Dichogamy is very much common in:
    a) Citrus fruits b) Annonaceous fruits
    c) Stone fruits d) Pome fruits
96. Leading raisin grape cultivar in the world is:
    a) Perlette b) Thompson Seedless
    c) Kishmish Chorni d) Riesling
97. Citrus having high malic acid content is:
    a) Acid lime b) Sweet lime
    c) Rangpur lime d) Tahiti lime
98. The pomegranate fruit 'Balausta' is a modified form of:
    a) Drupe b) Berry
    c) Sorosis d) Syconus
99. The headquarter of NBPGR (National Bureau of Plant Genetic Resources) is located at:
    a) Bangalore b) New Delhi
    c) Lucknow d) Jodhpur
100. Jackfruit (Artocarpus heterophyllus) is a native of:
    a) Bangladesh b) India
    c) Sri Lanka d) Indonesia
101. Mango malformation can be overcome by spraying NAA@:
    a) 100 ppm b) 200 ppm
    c) 300 ppm d) 400 ppm

102. Which of the following is a rich source of antioxidants?
   a) Guava   b) Black grape
   c) Avocado   d) Loquat

103. Seedlessness in guava is due to:
   a) Parthenocarpy   b) Stenospermocarpy
   c) Triploidy   d) Pollen sterility

104. Which of the following citrus variety remains green after harvesting even in the market?
   a) Kinnow   b) Mosambi
   c) Ponkan   d) Satgudi

105. Major hindrance in breeding of citrus fruits is:
   a) Long juvenile period   b) Nucellar embryony
   c) Self-incompatibility   d) Seedlessness

106. July Elberta is a variety of:
   a) Peach   b) Plum
   c) Nectarines   d) Pear

107. Most commonly used rootstock of Sapota is:
   a) Mahua   b) Khirni
   c) Adam apple   d) Mee tree

108. The crop in which pruning is done in summer:
   a) Phalsa   b) Fig
   c) Ber   d) Grape

109. FBD in pineapple takes place at:
   a) 30 leaves stage   b) 40 leaves stage
   c) 50 leaves stage   d) 60 leaves stage

110. Plumcot is an:
   a) Intergeneric hybrid   b) Interspecific hybrid
   c) Intervarietal hybrid   d) None of these

111. Which of the following techniques is used for detection of virus?
   a) ELISA   b) AFLP
   c) RFLP   d) RAPD

112. Fasciation in Sapota is caused by:
   a) Nutrient deficiency   b) Climatic factors
   c) Fungus   d) MLO's

113. Peach leaf curl disease is caused by:
   a) Virus b) Fungus
   c) Bacteria d) MLO's
114. Cock's comb is a disorder of:
   a) Cherry b) Avocado
   c) Loquat d) Sapota
115. Which of the following apricots gives the best dried product?
   a) Kaisha b) Halman
   c) Charmgaz d) Shakarpara
116. Which of the following fruit crops has the highest chromosome number?
   a) Mulberry b) Strawberry
   c) Cranberry d) Raspberry
117. Which of the following bacteria is primarily used for genetic transformation?
   a) *Agrobacterium tumefaciens* b) *Agrobacterium rhizogenes*
   c) *Agrobacterium radiobacter* d) *Pseudomonas fluorescens*
118. Avocado belongs to ....family:
   a) Moraceae b) Myrtaceae
   c) Rosaceae d) Lauraceae
119. Polymerase chain reaction (PCR) technique was invented by?
   a) Paul Berg b) Thomas Roderick
   c) Kary Mullis d) Botestein
120. Konkan Prolific is a variety of:
   a) Kokum b) Jackfruit
   c) Jamun d) Karonda
121. Which of the following mango varieties has highest number of perfect flowers?
   a) Langra b) Dashehari
   c) Bombay Green d) Pairi
122. The term 'Pitto' is associated with:
   a) Apple b) Pear
   c) Peach d) Plum
123. WTO is situated in:
   a) London b) Geneva
   c) Washington D.C. d) Paris

124. Which of the following is a naturally occurring inhibitor in plants?

a) 2,4-D b) BA

c) MH d) ABA

125. Which of the following plants is highly sensitive to salinity stress:

a) Banana b) Grape

c) Mango d) Citrus

Answers Key

1.	(a)	2.	(c)	3.	(d)	4.	(b)	5.	(d)	6.	(a)	7.	(a)
8.	(d)	9.	(b)	10.	(c)	11.	(b)	12.	(b)	13.	(a)	14.	(b)
15.	(b)	16.	(a)	17.	(d)	18.	(b)	19.	(d)	20.	(c)	21.	(b)
22.	(c)	23.	(b)	24.	(c)	25.	(b)	26.	(c)	27.	(d)	28.	(d)
29.	(a)	30.	(d)	31.	(a)	32.	(b)	33.	(b)	34.	(d)	35.	(a)
36.	(b)	37.	(b)	38.	(c)	39.	(c)	40.	(a)	41.	(b)	42.	(b)
43.	(c)	44.	(c)	45.	(c)	46.	(d)	47.	(d)	48.	(d)	49.	(b)
50.	(d)	51.	(c)	52.	(d)	53.	(d)	54.	(b)	55.	(c)	56.	(d)
57.	(d)	58.	(b)	59.	(a)	60.	(c)	61.	(b)	62.	(b)	63.	(d)
64.	(a)	65.	(a)	66.	(b)	67.	(b)	68.	(a)	69.	(d)	70.	(c)
71.	(d)	72.	(b)	73.	(a)	74.	(d)	75.	(b)	76.	(a)	77.	(d)
78.	(d)	79.	(b)	80.	(d)	81.	(d)	82.	(c)	83.	(a)	84.	(c)
85.	(b)	86.	(a)	87.	(a)	88.	(a)	89.	(a)	90.	(d)	91.	(d)
92.	(b)	93.	(b)	94.	(c)	95.	(b)	96.	(b)	97.	(b)	98.	(b)
99.	(b)	100.	(b)	101.	(b)	102.	(b)	103.	(c)	104.	(b)	105.	(b)
106.	(a)	107.	(b)	108.	(c)	109.	(b)	110.	(b)	111.	(a)	112.	(c)
113.	(b)	114.	(d)	115.	(b)	116.	(a)	117.	(a)	118.	(d)	119.	(c)
120.	(b)	121.	(a)	122.	(a)	123.	(b)	124.	(d)	125.	(d)		

7

ICAR – NET / ARS Fruit Science Exam – 2012

1. Mangosteen (*Garcinia mangostana*) belongs to the family:
   a) Anacardiaceae b) Caricaceae
   c) Rosaceae d) Clusiaceae
2. Amrit Sagar is a variety of:
   a) Aonla b) Bael
   c) Banana d) Grape
3. Nendran banana belongs to which genomic group?
   a) AAA b) AAB
   c) ABB d) AB
4. Which of the following fruit crop has maximum diversity in India?
   a) Apricot b) Olive
   c) Kiwifruit d) Peach
5. The most vigorous rootstock of grape is:
   a) Dogridge b) Salt Creek
   c) Harmony d) Freedom
6. Summer pruning is practiced in:
   a) Apple b) Cherry
   c) Peach d) Citrus
7. Dwarfness in papaya is controlled by:
   a) Recessive gene b) Additive gene
   c) Dominant gene d) Duplicate gene
8. Which of the following is an ornamental species of papaya?
   a) Vasconcellea candamarcensis b) Vasconcellea pentagona
   c) *Vasconcellea quercifolia* d) *Vasconcellea gracilis*

9. During VHT, the temperature attained by mango fruit pulp is:
   a) 43°C b) 45°C
   c) 47°C d) 50°C
10. Resistance to spongy tissue in mango is governed by:
   a) Dominant gene b) Recessive gene
   c) Additive gene d) Multiple genes
11. Preharvest fruit drop in citrus fruits can be controlled by the application of:
   a) GA3 b) Kinetin
   c) 2,4,5-T d) Ethrel
12. Which of the following amino acids is a precursor of auxin synthesis?
   a) Methionine b) Lysine
   c) Tryptophan d) Leucine
13. First successful transgenic fruit plant is produced in which fruit crop?
   a) Walnut b) Pear
   c) Papaya d) Plum
14. D-leaf is the best indicator of nutrient status of:
   a) Pineapple b) Grape
   c) Banana d) Apple
15. The nutrient element which is immobile in soil and hence applied only as a basal dose is:
   a) Ca b) N
   c) P d) Zn
16. The International Union for the Protection of New Varieties of Plants (UPOV) came into force in:
   a) 1961 b) 1968
   c) 1972 d) 1991
17. A major challenge in the protected cultivation of fruit crops is the increasing threat of:
   a) Whiteflies b) Aphids
   c) Fruit fly d) Nematodes
18. Expensive type of protected structure is:
   a) Poly house b) Glasshouse
   c) Polytunnel d) Phytotron

19. The dry neck is a physiological disorder of:
    a) Date palm b) Loquat
    c) Pineapple d) Avocado
20. Which fruit crop is most suitable for protected cultivation?
    a) Mango b) Apple
    c) Cherry d) Papaya
21. Which of the following citrus relatives is tetraploid in nature?
    a) Poncirus trifoliata b) Citrus latifolia
    c) Fortunella hindsii d) Eremocitrus glauca
22. Total number of AEZs in India are:
    a) 54 b) 60
    c) 65 d) 68
23. Which country is the largest producer of coconut in the world?
    a) Indonesia b) Philippines
    c) India d) Brazil
24. In rubber, the latex flow can be stimulated by the application of:
    a) Gibberellins b) Cytokinin
    c) Ethrel d) Auxin
25. Pusa Pratibha, a mango hybrid is a cross between:
    a) Amrapali × Sensation b) Dashehari ×Sensation
    c) Amrapali × Lal Sundari d) Amrapali × Vanraj
26. 'Pinching off' of terminal bud is recommended in the cultivation of which mango variety?
    a) Amrapali b) Mallika
    c) Arka Aruna d) Pusa Surya
27. Kiwifruit is a:
    a) Evergreen shrub b) Deciduous shrub
    c) Evergreen vine d) Deciduous vine
28. Scoring technique for banana classification was given by:
    a) Tanaka and Swingle b) Simmonds and Shepherd
    c) Kostermans and Bompard d) Bhattacharya and Dutta

29. Tangelo is a cross between:
    a) *Citrus reticulata* × *Citrus maxima*
    b) *Citrus reticulata* × *Citrus paradisi*
    c) *Citrus sinensis* × *Citrus maxima*
    d) *Citrus sinensis* × *Citrus paradise*
30. Kinnow mandarin is a:
    a) Intervarietal hybrid b) Interspecific hybrid
    c) Intergeneric hybrid d) Somatic hybrid
31. Maximum heating is obtained with which mulch material:
    a) Black mulch b) White mulch
    c) Straw d) Sawdust
32. Most widely used dwarfing rootstock of apple in the world is:
    a) M.9 b) M.27
    c) B.9 d) P.22
33. MM series of apple rootstock is resistant to :
    a) Virus b) Fire blight
    c) Woolly apple aphid d) San Jose Scale
34. Christ thorn is botanically known as:
    a) Capparis decidua b) Carissa congesta
    c) Cordia dichotoma d) Grewia asiatica
35. Alternate bearing is not common in:
    a) Apple b) Mango
    c) Tamarind d) Citrus
36. The main objective of precooling is:
    a) To remove field heat b) To fresh the product
    c) To improve fruit quality d) To inactivate enzymes
37. Flame grapefruit is a bud mutant of:
    a) Thompson b) Hudson
    c) Handerson d) Ruby Red
38. How many plants can be accommodated in high-density planting of pineapple?
    a) 10,000 – 20,000 b) 20,000 – 30,000
    c) 30,000 – 40,000 d) 40,000 – 50,000

39. In peach, the flowers are produced on:
   a) Current year growth b) Last year growth
   c) Spur d) Trunk
40. Which of the following is the highest salt tolerant fruit crop?
   a) Date palm b) Ber
   c) Aonla d) Guava
41. Strawberry is a:
   a) Short day plant b) Long day plant
   c) Day-neutral plant d) None of these
42. The yellow colour of papaya is due to the presence of:
   a) Carotene b) Caricaxanthin
   c) Quercetin d) Lycopene
43. The major hindrance in the breeding of banana is:
   a) Male sterility and dichogamy b) Female sterility and polyploidy
   c) Male sterility and pest incidence d) Male sterility and viral diseases
44. Degreening is not applicable in:
   a) Mango b) Banana
   c) Citrus d) Guava
45. Defoliation in Annona results in a good crop. It is achieved by the spray of 0.05%:
   a) KSO4 b) KI
   c) KIO3 d) KCI
46. Which fruit crop doesn't have a developed rootstock till date?
   a) Cherry b) Pear
   c) Apple d) Litchi
47. Litchi is commercially propagated by:
   a) T - budding b) Wedge – grafting
   c) Air layering d) Hardwood stem cutting
48. Which fruit crop is mainly grown in polytunnel (protected condition) in India?
   a) Papaya b) Banana
   c) Strawberry d) Pineapple

49. The Botanical name of Mountain papaya is:
    a) *Vasconcellea parviflora* L.
    b) *Vasconcellea candamarcensis* Hooker
    c) *Vasconcellea papaya* L.
    d) *Vasconcellea microcarpa* Jack
50. A grape cultivar whose genome has been completely sequenced:
    a) White Riesling b) Chardonnay
    c) Red Globe d) Pinot Noir
51. 0.1 g of a compound mixed in 10 litres of water gives a solution of:
    a) 10 ppm b) 100 ppm
    c) 1000 ppm d) 10,000 ppm
52. Which of the following is a deciduous species of citrus?
    a) Fortunella hindsii b) Citrus unshiu
    c) Poncirus trifoliata d) Microcitrus australis
53. Which of the following species of papaya is resistant to papaya ringspot virus?
    a) Vasconcellea parviflora b) Vasconcellea microcarpa
    c) Vasconcellea cauliflora d) Vasconcellea pentagona
54. Severe pruning is practiced in:
    a) Aonla b) Ber
    c) Jamun d) Pomegranate
55. The hybrid formed by fusion of two somatic cells is known as:
    a) Cybrid b) Gametic hybrid
    c) Somatic hybrid d) Distant hybrid
56. Natural fruit set in custard apple is:
    a) 2-3% b) 5-6%
    c) 7-8% d) 9-10%
57. Wood apple is propagated by:
    a) Cutting b) Budding
    c) Grafting d) Layering
58. In India, tissue culture has been commercially exploited in:
    a) Papaya b) Grape
    c) Banana d) Pineapple

59. Which of the following is a dwarfing rootstock of a mango?
    a) Olur b) Moovadan
    c) Totapuri Red Small d) Kurakkan
60. The compound present in grapefruit which is helpful for people suffering from diabetes is:
    a) Limonin b) Naringin
    c) Lycopene d) Nomilin
61. Myrobalan B is an important rootstock of:
    a) Cherry b) Peach
    c) Plum d) Almond
62. Which of the following is an ethylene absorbent?
    a) KMnO4 b) KNO3
    c) K2SO4 d) KCI
63. The pomegranate fruit 'Balausta' is a modified form of:
    a) Drupe b) Berry
    c) Nut d) Sorosis
64. Which of the following citrus fruits contains malic acid?
    a) Mexican lime b) Lemon
    c) Pummelo d) Sweet lime
65. Under Maharashtra condition, the grape is pruned:
    a) Once a year b) Twice a year
    c) Thrice a year d) Round the year
66. Papaya is native of:
    a) Northern Mexico b) Sothern Mexico
    c) Eastern Mexico d) Western Mexico
67. The enzyme bromelain is extracted from:
    a) Bael b) Fig
    c) Jamun d) Pineapple
68. Fuerte is an important variety of:
    a) Apple b) Avocado
    c) Papaya d) Persimmon
69. Secondary centre of origin of Banana is:
    a) Malaysia b) New Guinea
    c) India d) Bangladesh

70. The intervening period between the sequential emergences of leaves on the main stem of a plant is known as:

a) Phyllochron b) Cytochrome

c) Helicoid d) Phytotron

71. Granulation is a physiological disorder of:

a) Grape b) Citrus

c) Litchi d) Pomegranate

72. Mango seed is classified as:

a) Orthodox b) Recalcitrant

c) Intermediate d) None of these

73. The bud wood of temperate plant species is stored at

a) 32°F b) 48°F

c) 72°F d) 96°F

74. Which of the following is a seedless variety of mango?

a) Ratna b) Sindhu

c) Konkan Ruchi d) Alphonso

75. Cell division continues till maturity in fruit development of:

a) Citrus b) Strawberry

c) Persimmon d) Avocado

76. In HTST, the fruit juices are pasteurized at:

a) 72°C for 15 seconds b) 80°C for 30 seconds

c) 85°C for 30 minutes d) 100°C for 10 minutes

77. Dichogamy is very common in:

a) Citrus fruits b) Annonaceous fruits

c) Pome fruits d) All of these

78. The sex form of Pusa Majesty and Pusa Delicious varieties of papaya is:

a) Monoecious b) Dioecious

c) Gynodioecious d) Hermaphrodite

79. Spacing followed in high density planting of mango is:

a) 1.5 × 1.5 m b) 2.5 × 2.5 m

c) 3.5 × 3.5 m d) 5.5 × 5.5 m

80. Kew variety of pineapple belongs to the group:

a) Queen b) Spanish

c) Cayenne d) Abcaxi

81. Which of the following grape varieties sets fruits by stimulative parthenocarpy?

a) Beauty Seedless b) Black Corinth
c) Perlette d) Black Muscat

82. Gene of resistance for mango malformation is found in:

a) Bombay Green b) Bhadauran
c) Langra d) Neelum

83. Chemical mutagen commonly used in mutation breeding is:

a) N-Nitroso-N-Methylurea b) N-Nitroso-N-Methyl-Urethane
c) Ethyl Methane Sulphonate d) None of these

84. Cultivation of transgenic papaya has been commercialized in:

a) Hawaii b) China
c) Japan d) Canada

85. Post-harvest berry drop in grape can be controlled by the spray which of the following one week before harvest:

a) Ethephone @25 ppm b) 2,4-D @25 ppm
c) NAA @50 ppm d) GA3 @50 ppm

86. Transgraft has:

a) Transgenic rootstock b) Transgenic scion
c) Both are transgenic d) None of these

87. In mango, only …… hermaphrodite flowers develop fruit to maturity.

a) <0.1% b) <1%
c) <5% d) <10%

88. Molecular marker used for physical mapping of gene is:

a) RFLP b) RAPD
c) AFLP d) ISSR

89. Zygodormancy is found in:

a) Aonla b) Bael
c) Ber d) Citrus

90. Coffee is a native of:

a) Ethiopia b) Brazil
c) India d) China

91. The following hormone is present naturally in the ovaries of the parthenocarpic fruits:

a) Auxin	b) Cytokinin
c) Gibberellin	d) Ethylene

92. Marmalade is prepared from:

a) Lime	b) Orange
c) Pummelo	d) Grapefruit

93. The main photosynthetic product in Rosaceous fruit crops is:

a) Sorbitol	b) Mannitol
c) Fructose	d) Sucrose

94. The ability of a plant cell to develop into the whole plant is known as:

a) Prepotency	b) Multipotency
c) Totipotency	d) Pluripotency

95. IPR stands for:

a) Indigenous Property Right	b) Intellectual Property Right
c) Indian Property Right	d) Individual Property Right

96. Cryopreservation of pollen is done in liquid nitrogen at

a) 196°C	b) -196°C
c) 0°C	d) 60°C

97. The best fruit recommended for diabetes is:

a) Mango	b) Avocado
c) Sitaphal	d) Jamun

98. The crop which does not belong to the same family:

a) Mango	b) Pecan nut
c) Cashew nut	d) Pistachio nut

99. Red apple varieties are the results of:

a) Periclinal chimeras	b) Sectorial chimeras
c) Mericlinal chimeras	d) Normal chimeras

100. The ideal weight of banana suckers for propagation is:

a) 0.25-0.75 kg	b) 0.75-1.0 kg
c) 1.5-2.0 kg	d) 2.0-2.5 kg

101. The induction flowering by exposure to the chilling treatment is known as:
   a) Stratification b) Scarification
   c) Vernalization d) Photoperiodic

102. The PGR used as a substitute for a chilling requirement in many temperate fruits is:
   a) Ethrel b) Kinetin
   c) NAA d) GA3

103. The best time of budding in ber is:
   a) July - August b) February - March
   c) September - October d) May – June

104. Botanically, Sapota fruit is a:
   a) Drupe b) Berry
   c) Nut d) Syconus

105. Which of the following is an ultra-dwarfing rootstock for citrus?
   a) Troyer citrange b) Rangpur Lime
   c) Flying Dragon d) Sour orange

106. The inflorescence of cashew nut is termed as:
   a) Andromonoecious b) Gynoecious
   c) Polygamy gynodioecious d) None of these

107. The natural cytokine 'zeatin' was first extracted from:
   a) Maize b) Sorghum
   a) Rice d) Barley

108. Which rootstock of citrus is resistant to Tristeza virus?
   a) Sour orange b) Rangpur lime
   c) Trifoliate orange d) Rough lemon

109. A leading raisin grape cultivar is:
   a) Perlette b) Flame Seedless
   c) Thompson Seedless d) Riesling

110. The main insect reported to be responsible for affecting pollination in mango is:
   a) Honey bee b) Housefly
   c) Fruit fly d) Wasp

111. Pollination of fig flower by the Blastophaga wasps takes place through:

a) Fascination b) Parthenogenesis
c) Caprification d) Apomixis

112. Which type of flowers is contained in the mango inflorescence?

a) Male and hermaphrodite b) Male and female
c) Male and neutral d) Female and hermaphrodite

113. Burr knots in apple is related to:

a) Clonal rootstock b) Woolly apple aphid
c) Apple scab d) Cuttings

114. Growth regulator CCC is used is grape for:

a) Increase vegetative growth b) Increase TSS
c) Increase fruitfulness d) None of these

115. AVG is an inhibitor of:

a) Ethylene biosynthesis b) Gibberellin biosynthesis
c) Auxin biosynthesis d) Cytokinin biosynthesis

116. Highest papain yielding variety of papaya is:

a) Pusa Delicious b) Pusa Majesty
c) CO-2 d) CO-5

117. Which of the following is a self-fertile variety of apple?

a) Baldwin b) Mutsu
c) Golden Delicious d) Winesap

118. A natural growth inhibitor present in plants is:

a) MH b) ABA
c) GA_3 d) CCC

119. Harvest index can be calculated by using the formula:

a) Economic yield / Biological yield
b) Biological yield / Economic yield
c) Leaf yield/fruit yield
d) None of these

120. Chimaeras can be maintained through:

a) Sexual propagation b) Asexual propagation
c) Hybridization d) Mutation

121. Ultra dwarf rootstock of apple is:
   a) M 9 b) M 27
   c) P 22 d) MM 104
122. A one-year-old shoot of grape in which the fruit bunches appear is known as:
   a) Arms b) Canes
   c) Shoots d) Spurs
123. The skin colour in mango is governed by:
   a) Polygenes b) Oligo genes
   c) Additive genes d) Dominant genes
124. Which of the following is an ornithophilous fruit crop?
   a) Pineapple b) Banana
   c) Strawberry d) Oil palm
125. Cauliflory is observed in:
   a) Jackfruit b) Grape
   c) Mango d) Citrus
126. Carambola is a rich source of:
   a) Propanoic acid b) Malic acid
   c) Oxalic acid d) Linoleic acid
127. A banana cultivar, highly susceptible to panama wilt is:
   a) Dwarf Cavendish b) Gross Michel
   c) Grand Nain d) Pisang Lilin
128. Pineapple is native to:
   a) Brazil b) Paraguay
   c) Peru d) Mexico
129. Most commonly used dryer for drying of fruit juices is:
   a) Kiln dryer b) Cabinet dryer
   c) Rotary dryer d) Spray dyer
130. Which of the following is the richest source of vitamin C:
   a) Aonla b) Guava
   c) Citrus d) Barbados cherry
131. Based on the mode of pollination, tall coconuts are:
   a) Self-pollinated b) Cross-pollinated
   c) Often cross-pollinated d) All of these

132. In papaya, female plants can be produced in bulk through:
   a) Anther culture b) Ovule culture
   c) Meristem culture d) Cell culture

133. Meadow orchard in guava is based on the principle of:
   a) Regular topping and hedging of plants
   b) Use of dwarfing rootstocks
   c) Use of plant growth regulators to reduce the height
   d) All of these

134. Most commonly used training system for most of the fruit crops is:
   a) Central leader b) Open center
   c) Modified central leader d) Tatura trellis

135. Which of the following is a mutant of Thompson Seedless?
   a) Sharad Seedless b) Tas-e-Ganesh
   c) Flame Seedless d) Bangalore Blue

136. Kaveri is an improved variety of:
   a) Coffee b) Persimmon
   c) Cherry d) Passion fruit

137. Which of the following is known as the 'Miracle fruit of China'?
   a) Litchi b) Loquat
   c) Persimmon d) Kiwifruit

138. Dormex (Hydrogen cyanamide) is used to hasten early bud break-in:
   a) Litchi b) Papaya
   c) Apple d) Grape

139. Santa Rosa is a cultivar of:
   a) European plum b) Japanese plum
   c) American plum d) Canadian plum

140. Across the world, grapes are mostly processed into:
   a) Raisin b) Wine
   c) Juice d) Canning

141. Which of the following is monocotyledonous, monoecious, monocarpic, mesophytes, herbaceous plant?
   a) Pineapple b) Papaya
   c) Litchi d) Banana

142. The plant growth regulator used for increasing berry size of grapes is:

a) GA_3 b) ABA

c) Kinetin d) NAA

143. The best irrigation system for protected cultivation is:

a) Sprinkler system b) Drip system

c) Basin system d) Flood system

144. Strawberry is commercially propagated by:

a) Suckers b) Runners

c) Offsets d) Stolen

145. A suitable structure for plant growth studies is:

a) Shed net house b) Low cost green house

c) Poly tunnels d) Phytotron

146. Hen and Chicken disorder is found in:

a) Grape b) Papaya

c) Peach d) Date palm

147. Indian wild strawberry is botanically known as:

a) Fragaria vesca b) Fragaria × ananassa

c) *Fragaria virginiana* d) *Fragaria indica*

148. The most profitable intercrop in the banana orchard is:

a) Papaya b) Guava

c) Turmeric d) Ginger

149. The technique used for producing virus-free plants in citrus is known as:

a) Micro budding b) Micro grafting

c) Epicotyl grafting d) All of these

150. Cavendish banana belongs to which of the following groups:

a) AAA b) AAB

c) ABB d) AB

Answers Key

1.	(d)	2.	(c)	3.	(b)	4.	(a)	5.	(a)	6.	(a)	7.	(a)
8.	(d)	9.	(c)	10.	(b)	11.	(c)	12.	(c)	13.	(a)	14.	(a)
15.	(c)	16.	(b)	17.	(d)	18.	(d)	19.	(d)	20.	(d)	21.	(c)
22.	(b)	23.	(a)	24.	(c)	25.	(a)	26.	(a)	27.	(d)	28.	(b)
29.	(b)	30.	(b)	31.	(a)	32.	(a)	33.	(c)	34.	(b)	35.	(d)
36.	(a)	37.	(c)	38.	(d)	39.	(b)	40.	(a)	41.	(a)	42.	(b)
43.	(b)	44.	(d)	45.	(b)	46.	(d)	47.	(c)	48.	(c)	49.	(b)
50.	(d)	51.	(a)	52.	(c)	53.	(c)	54.	(b)	55.	(c)	56.	(a)
57.	(b)	58.	(c)	59.	(c)	60.	(b)	61.	(c)	62.	(a)	63.	(b)
64.	(d)	65.	(b)	66.	(b)	67.	(d)	68.	(b)	69.	(c)	70.	(a)
71.	(b)	72.	(b)	73.	(a)	74.	(b)	75.	(d)	76.	(c)	77.	(b)
78.	(c)	79.	(b)	80.	(c)	81.	(b)	82.	(b)	83.	(c)	84.	(a)
85.	(c)	86.	(a)	87.	(a)	88.	(a)	89.	(a)	90.	(a)	91.	(c)
92.	(b)	93.	(a)	94.	(c)	95.	(b)	96.	(a)	97.	(d)	98.	(b)
99.	(a)	100.	(b)	101.	(c)	102.	(d)	103.	(a)	104.	(b)	105.	(c)
106.	(a)	107.	(a)	108.	(c)	109.	(c)	110.	(b)	111.	(c)	112.	(a)
113.	(a)	114.	(c)	115.	(a)	116.	(d)	117.	(c)	118.	(b)	119.	(a)
120.	(b)	121.	(b)	122.	(b)	123.	(a)	124.	(b)	125.	(a)	126.	(c)
127.	(b)	128.	(b)	129.	(b)	130.	(d)	131.	(b)	132.	(b)	133.	(a)
134.	(c)	135.	(b)	136.	(d)	137.	(d)	138.	(d)	139.	(b)	140.	(b)
141.	(d)	142.	(a)	143.	(b)	144.	(b)	145.	(d)	146.	(a)	147.	(a)
148.	(d)	149.	(b)	150.	(a)								

8

ICAR – NET / ARS Fruit Science Exam – 2013

1. Pineapple is commercially propagated by:
   a) Slips b) Suckers
   c) Crown d) Runners
2. Colt is a dwarfing rootstock of:
   a) Cherry b) Peach
   c) Plum d) Apricot
3. Palmette system of training is a modification of:
   a) Central leader b) Spindle bush
   c) Dwarf pyramid d) Open center
4. Vikram variety of acid lime was released from:
   a) MPKV, Rahuri b) KKV, Dapoli
   c) MAU, Parbhani d) PDKV, Akola
5. Pusa Nanha a dwarf mutant of papaya was developed using:
   a) X-rays b) Thermal neutrons
   c) þ-rays d) Gamma rays
6. Which of the following is a dwarfing rootstock of plum?
   a) Grand Ferrade b) St. Julian
   c) Siberian - C d) Pixy
7. Multiple hedge row system is widely practiced in:
   a) Guava b) Peach
   c) Apple d) Cherry
8. Sunehari, a hybrid variety of apple, is a cross between :
   a) Golden Delicious x Ambri b) Ambri x Golden Delicious
   c) Golden Delicious x Lal Ambri d) Ambri x Granny Smith

9. Plastic material used for protected cultivation should be:
   a) UV stabilized b) 200 micron thickness
   c) Low cost d) All of these
10. Resistance to spongy tissue in .... is governed by:
   a) Dominant gene b) Recessive gene
   c) Additive gene d) Multiple gene
11. The chamber temperature maintained in vapour heat treatment of mango is 50-52°C,while the temperature of mango pulp is:
   a) 45°C b) 46°C
   c) 47°C d) 48°C
12. Blossom end rot of grape is caused due to the deficiency of:
   a) Mg b) B
   c) Ca d) Fe
13. The Ca deficiency symptoms in plants first occur at:
   a) Older leaves b) Younger leaves
   c) Terminal bud d) None of the above
14. The secondary place of origin of mango is:
   a) Myanmar b) Western Malaysia
   c) Florida d) New Guinea
15. The conservation of fruit crops germplasm outside their natural habitat is known as:
   a) *Ex-situ* conservation b) *In-situ* conservation
   c) On farm conservation d) None of the above
16. Alcohol content in apple cider is:
   a) 4-6% b) 6-8%
   c) 8-10% d) 10-12%
17. Dwarf pyramid system of training is a modification of:
   a) Spindle bush b) Lincoln canopy
   c) Cordon d) Paimette
18. In which of the following systems, every tree has only two limbs growing east and west at an angle of 60° to horizontal?
   a) Cordon b) Tatura trellis
   c) Spindle bush d) Solen system

19. Pusa Urvashi is a cross between:
    a) Hur × Bharat Early
    b) Hur × Beauty Seedless
    c) Madeleine Angevine × Ruby Red
    d) Queen of Vineyard × Black Kishmish
20. Arka Aruna, a mango hybrid, is a cross between:
    a) Alphonso × Banaganapalli b) Banaganapalli × Alphonso
    c) Alphonso × Janardan Pasand d) Alphonso × Neelum
21. Which of the following species of *Ziziphus* has originated from China?
    a) *Ziziphus apetala* b) *Ziziphus incurva*
    c) *Ziziphus jujuba* d) All the above
22. Irradiation of mango fruit is recommended for protection against:
    a) Stone weevil b) Fruit fly
    c) Mealy bug d) All the above
23. Which of the following is not associated with apple production?
    a) Premature leaf fall b) Replant problem
    c) Ring spot disease d) Root rot
24. How much amount of GA3 is required for making a solution of 200 ppm in 1000 ml water?
    a) 2 mg b) 20 mg
    c) 200 mg d) 2g
25. Which of the following series of apple rootstocks is resistant to viruses?
    a) MI series b) MM series
    c) EMLA series d) All of these
26. Apomixes is a characteristics feature of *Malus* species.
    a) Obligate b) Facultative
    c) Non recurrent d) Nucellus
27. Grape rootstock resistant to nematode and salt is:
    a) Dogridge b) Riparia Glorie
    c) St. George d) Temple
28. Kniffin system of grape training was developed by Mr. William Kniffin of:
    a) France b) Germany
    c) New York d) New Zealand

29. Which of the following is considered a 'Dollar earning crop'?
    a) Walnut b) Cashew nut
    c) Pecan nut d) Hazel nut
30. Arka Amulya, a hybrid variety of guava, is a cross between:
    a) Allahabad Safeda × Seedless
    b) Seedless × Allahabad Safeda
    c) Red Fleshed × Saharanpur Seedless
    d) Banarasi × Allahabad Safeda
31. Which of the following is a hybrid variety of pomegranate?
    a) Ganesh b) Mridula
    c) G-137 d) Jyoti
32. Which of the following is/are polyembryonic species of citrus?
    a) *Citrus reticulata* b) *Citrus paradisi*
    c) *Citrus jamhhiri* d) All of these
33. Citrange is a cross between:
    a) C. sinensis × Poncirus trifoliata
    b) C. paradisi × Poncirus trifoliata
    c) C. reticulata × Poncirus trifoliata
    d) C. aurantium × Poncirus trifoliata
34. A fruit crop cultivated worldwide under protected structure is:
    a) Grape b) Papaya
    c) Pineapple d) Strawberry
35. Which of the following fruit crops is mostly stored under cold storage conditions in India?
    a) Apple b) Mango
    c) Peaches d) Banana
36. Water saving percentage in banana due to drip irrigation is:
    a) 20-30% b) 30-40%
    c) 40-50% d) 50-60%
37. Protoplast fusion/parasexual hybridization is used to produce:
    a) Cybrids b) Hybrids
    c) Soma clones d) None of these

38. The most stable sex form in papaya is:
    a) Male b) Female
    c) Hermaphrodite d) All of these
39. The cross between Euvitis and Muscadinia is:
    a) Fertile b) Sterile
    c) Do not set fruits d) None of these
40. Fruit size in mango is governed by:
    a) Single gene b) Poly genes
    c) Additive genes d) Complementary genes
41. Which of the following is not a method of genetic transformation?
    a) X-ray mediated
    b) Agrobacterium mediated
    c) Microprojectile bombardment based
    d) Liposome mediated
42. Which of the following is a transgenic variety of papaya?
    a) Sinta b) Hortus Gold
    c) Red Lady d) Rainbow
43. High Gate is a semi dwarf mutant of:
    a) Dwarf Cavendish b) Gross Michel
    c) Grand Naine d) Saba
44. Genomic constitution of FHIA-1 (Gold finger) is:
    a) AAAA b) ABBB
    c) AAAB d) ABB
45. Red skin colour in litchi is due to the presence of:
    a) Lycopene b) Prolycopene
    c) Quercetin d) Anthocyanin
46. Which of the following is a climacteric fruit?
    a) Bael b) Avocado
    c) Apple d) Litchi
47. Denavelling is done in:
    a) Papaya b) Banana
    c) Apple d) Pineapple

48. Florina, a scab resistant variety of apple was introduced in India from:

a) Florida
b) France
c) USA
d) England

49. Protogynous diurnally synchronous dichogamy (PDSD) flowering habit is a rule in:

a) Chestnut
b) Pecan nut
c) Custard apple
d) Avocado

50. Pollination in pecan nut is facilitate by:

a) Bees
b) Wind
c) Wasp
d) Birds

51. Kagzi lime (*Citrus aurantifolia*) is an indicator plant of:

a) Citrus greening
b) Citrus tristeza virus
c) Citrus psorosis
d) Citrus exocortis

52. Pineapple is the native of:

a) Brazil
b) Paraguay
c) Peru
d) Mexico

53. In mango, regularity in bearing to some extent can be induced with the application of:

a) ABA
b) Auxin
c) Paclobutrazol
d) Ethephon

54. 'Goma Priyanka' is an improved variety of:

a) Aonia
b) Ber
c) Jamun
d) Pomegranate

55. The movement of auxin in stem (aerial) parts of plants is:

a) Acropetal
b) Basipetal
c) Non-directional
d) None of these

56. The scientific name of Mountain papaya is:

a) *Vasconcella parviflora* L.
b) *Vasconcella candamarcensis* Hooker
c) *Vasconcella papaya* L.
d) *Vasconcella microcarpa* Jack

57. Coat protein mediated resistance has been developed in:
    a) Papaya b) Banana
    c) Citrus d) All of these
58. Based on photoperiodic requirement, apple is a classified as:
    a) Short day plant b) Long day plant
    c) Day neutral plant d) None of the above
59. Which chemical is used for breaking bud dormancy in grape?
    a) Hydrogen cyanamide b) KNO3
    c) Hydrogen peroxide d) All of these
60. The colour breaking stage in grape is termed as:
    a) Breakeven point b) Verasion
    c) Teinturier d) Turning stage
61. The leading state in litchi production in India is:
    a) Punjab b) Bihar
    c) Maharashtra d) West Bengal
62. Which of the following is an ethylene absorbent?
    a) KMnO4 b) KNO3
    c) K2SO4 d) KCl
63. Best time of pruning in ber in north India is:
    a) March-April b) April-May
    c) May-June d) December-January
64. Most of the aonla (*Emblica officinalis*) cultivars are released from:
    a) NDUAT, Ayodhya b) CISH, Lucknow
    c) MAU, Parbhani d) CSAUAT, Kanpur
65. Which of the following fruit crops does not belong to the family Anacardiaceae?
    a) Mango b) Cashew nut
    c) Pecan nut d) Pistachio nuts
66. Which one of the following is the cause of self-unfruitful in loquat?
    a) Gametophytic incompatibility b) Pollen sterility
    c) Ovule abortion d) Cross incompatibility
67. Zero energy cool chamber operates on the principle of:
    a) Law of thermodynamics b) Evaporative cooling
    c) Charles's law d) Boyle's law

68. Waxing of fruits is done to reduce:
    a) Respiration b) Transpiration
    c) Both d) None of these
69. The enzyme responsible for converting pectin into pectic acid is:
    a) Polygalacturonase b) Protopectinase
    c) Pectic methyl esterase d) Pectinase
70. Quince C is a dwarfing rootstock of:
    a) Apple b) Pear
    c) Peach d) Cherry
71. Important constituent of jelly is:
    a) Sugar b) Acid
    c) Water d) Pectin
72. Choke throat in Gross Michael banana is due to:
    a) Low temperature b) Low humidity
    c) High temperature h) High humidity
73. Low temperature treatment of seeds to promote germination is known as:
    a) Scarification b) Stratification
    c) Seed priming d) None of the above
74. Fuzzy nectarines are also called :
    a) Dried plums b) Peaches
    c) Cherries d) Dried apricots
75. Dwarfness in banana is controlled by:
    a) Dominant gene b) Recessive gene
    c) Duplicate gene d) Co-dominant gene
76. Skin colour in mango is controlled by:
    a) Dominant gene b) Recessive gene
    c) Poly genes d) Additive genes
77. Haploid chromosome number (X=) of apple is:
    a) 14 b) 17
    c) 28 d) 34
78. Which of the following rootstocks is resistant to fire blight of apple?
    a) M9 b) EMLA9
    c) MM106 d) Ottawa

79. Under UPOV system, duration of protection for tree and vine crops is:
    a) 15 year b) 18 year
    c) 20 year d) 25 year
80. Training system followed in pomegranate is:
    a) Single stem training b) Multi stem training
    c) Modified central leader d) None of these
81. Double sigmoid growth curve is observed in:
    a) Mango b) Apple
    c) Fig d) Citrus
82. Conservation of propagules is done in liquid N2 at a temperature of:
    a) 0°C b) 196°C
    c) -196°C d) -60°C
83. Allopolyploid cultivar of mango is:
    a) Bappakai b) Vellaikolamban
    c) Goa d) Olur
84. A nutrient element which improves the quality of fruit is:
    a) N b) P
    c) K d) Ca
85. The most suitable C/N ratio for growth and fruiting is:
    a) CC/NNN b) C/NNN
    c) CCC/NN d) CCCC/N
86. Which fruit is the richest source of iron?
    a) Date palm b) Phalsa
    c) Karonda d) Apple
87. In mango, only.......................hermaphrodite flowers develop to maturity.
    a) <0.1% b) <1%
    c) <5% d) <10%
88. Which fruit gives highest calories per 100 g of edible portion?
    a) Banana b) Cashew nut
    c) Almond d) Apple
89. Which of the following compounds is responsible for aroma in over- ripe banana?
    a) Hexanal b) Isopentanol
    c) Eugenol d) Methyl salicylate

90. When *Ziziphus mauritiana* is budded onto *Ziziphus nummularia*, it imparts:
    a) Disease resistance b) Dwarfness
    c) Drought tolerance d) Salt tolerance

91. Most of the mandarin cultivation has been diversified in:
    a) North-eastern mountainous region
    b) North-western mountainous region
    c) South-eastern mountainous region
    d) South-western mountainous region

92. Best Maturity Index for mandarin is:
    a) Oil content b) Acidity content
    c) Juice content d) TSS/acid ratio

93. Which of the following apple cultivars does not belong to India?
    a) Shreen b) Firdaus
    c) Akbar d) Gloster

94. The ratio of edible and non-edible matters in an ideal variety of mango should be:
    a) 1.0 to 2.0 b) 2.0 to 3.0
    c) 3.0 to 4.0 d) 4.0 to 5.0

95. *Mangifera cochinchinensis* occurs in:
    a) China b) Thailand
    c) Philippines d) Vietnam

96. A mango species which is hardly known outside its native region is:
    a) *Mangifera merilli* b) *Mangifera pajang*
    c) *Mangifera minniifolia* d) *Mangifera magnifica*

97. An Apple cultivar having high export quality in the world is:
    a) Red Delicious b) Golden Delicious
    c) Pink Lady d) Granny Smith

98. Removal of suckers is commonly practiced in:
    a) Chestnut b) Pecan nut
    c) Hazelnut d) Pistachio nut

99. The pH of an ideal rooting medium should be:
    a) 4.5-5.5 b) 5.5-6.5
    c) 6.5-7.5 d) 7.5-8.5

100. Bunch covering in banana increases the yield up to:

a) 10-15% b) 15-20%

c) 20-25% d) 25-30%

101. The size of the CFB carton accommodating 4 kg grapes is:

a) 25.5 × 18.5 × 11.5 cm b) 37 × 25.5 × 11.5 cm

c) 50 × 30 × 12 cm d) 40 × 20.5 × 10 cm

102. The ideal centre height of greenhouse for mango and papaya should he:

a) 2.5-3.5 m b) 3.5 to 4.5 m

c) 4.5 to 5.5 in d) 5.5 to 6.5 m

103. Which type of plum requires pinning and thinning?

a) European plum b) Japanese plum

c) American plum d) All of these

104. Most of the pineapple varieties take .... from planting to harvesting ?

a) 9-12 months b) 12-15 months

c) 15-18 months d) 18-21 months

105. In fertigation, fertilizers can be scheduled based on:

a) Tissue nutrient status b) Soil nutrient status

c) Visual deficiency symptoms d) None of these

106. Most commonly used explant for pineapple is:

a) Axillary buds b) Leaf base

c) Slips d) Vegetative bud

107. Saline soils are formed in:

a) High rainfall areas b) Hilly areas

c) Near to coastal regions d) Peninsular regions

108. In India, high density planting is commercially established in:

a) Mango b) Papaya

c) Citrus d) Guava

109. The maximum amount of SO2 (KMS) allowed in preservation of fruit juice is:

a) 100 ppm b) 350 ppm

c) 600 ppm d) 700 ppm

110. The preservative action of sugar is due to:
   a) Antioxidant b) Osmosis
   c) Inactivating enzymes d) Creating anaerobic condition
111. Haploid plantlets are developed from:
   a) Embry culture b) Anther culture
   c) Meristem culture d) Somatic hybridization
112. Molecular markers can be used to:
   a) Identify clones and cultivars b) To study diversity analysis
   c) Phylogenetic studies of species d) All of these
113. The genetic variability present among the tissue cultured cells is referred to as:
   a) Soma-clonal variation b) Mutation
   c) Chimeras d) None of these
114. In dark reaction of photosynthesis, CO2 ends in:
   a) ATP b) NADPH
   c) Sugars d) Water
115. Gene expression of ripening is not characterized in:
   a) Mango b) Banana
   c) Papaya d) Litchi
116. Which of the following temperate fruit requires frequent pruning?
   a) Apple b) Pear
   c) Japanese plum d) Cherry
117. 0.5 g of salt in 5 litre of water gives a solution of:
   a) 250 ppm b) 500 ppm
   c) 2500 ppm d) 25000 ppm
118. Which of the following is an organic phosphate solving bio-fertilizer?
   a) VAM b) Azola
   c) Rhizobium d) Azotobacter
119. Cultivated grape (*Vitis vinifera*) is the hybrid between:
   a) *Vitis parviflora* × *Vitis vulpina*
   b) *Vitis vulpina* × *Vitis labrusca*
   c) *Vitis lanata* × *Vitis vulpina*
   d) *Vitis parviflora* × *Vitis labrusca*

120. Water saving percentage in banana due to drip .... Repeat question .... irrigation is:
   a) 20-30% b) 30-40%
   c) 40-50% d) 50-60%

121. Which of the following is not an essential requirement for protection of plant varieties?
   a) Distinctness b) Uniformity
   c) Stability d) Examination

122. Which of the following methods is used for preservation of germplasm of fruit crops:
   a) *In-situ* conservation b) *Ex-situ* conservation
   c) On farm conservation d) None of these

123. Which of the following plant growth regulators is used to induce rooting in cutting?
   a) IAA b) NAA
   c) IBA d) 2-4 D

124. Which of the following groups of pineapple is not commercially cultivated?
   a) Pandora b) Abacaxi
   c) Queen d) Cayenne

125. Spongy tissue in mango is a:
   a) Fungal disease b) Nutritional disorder
   c) Physiological disorder d) Bacterial disease

126. Annona atemoya is a cross between:
   a) *A. squamosa* × *A. cherimola* b) *A. cherimola* × *A. reticulata*
   c) *A. squamosa* × *A. muricata* d) *A. squamosa* × *A. reticulata*

127. Alternate bearing in mango is caused due to:
   a) Climatological factors b) Inadequate C/N ratio
   c) Age and vigour of the tree d) All of these

128. Head space leaving in canning of fruits is:
   a) 0.1-0.3 cm b) 0.3-0.5 cm
   c) 0.5-0.7 cm d) 0.7-0.9 cm

129. The most commonly used dryer for drying of fruit juices is:
   a) Kiln dryer b) Cabinet dryer
   c) Rotary dryer d) Spray dyer

130. Which of the following practice is employed for converting old, senile and unproductive orchards into productive ones?

a) Frame working   b) Top working
c) Double working   d) None of these

131. Strawberry cultivation is not suitable under:

a) Hilly regions   b) North Indian plains
c) Undulated hilly regions   d) Protected cultivation

132. Recalcitrance of seeds is considered to be a major problem in seed storage of:

a) Apple   b) Grape
c) Date palm   d) Papaya

133. Which of the following statement is not true in case of sigatoka leaf spot of banana:

a) It was first recognized in Fiji
b) All AAA clones are susceptible while ABB are resistant
c) Caused by fungus *Mycosphaerella fijiensis*
d) Gross Michael is highly resistant to this disease

134. Which of the following problems is a not related to apple cultivation?

a) Scab   b) White root rot
c) Silver leaf canker   d) Ring spot disease

135. Best covering material for soil solarization is:

a) Black polythene   b) White polythene
c) Transparent polythene   d) None of these

136. Vapour heat treatment is recommended for disinfection of mango against:

a) Fruit fly   b) Stone weevil
c) Both fruit fly and stone weevil   d) White fly

137. Papaya is cultivated under greenhouse to provide protection against:

a) High wind velocity   b) Papaya ring spot virus
c) Frost   d) All of these

138. Which aspect is not related to crop regulation?

a) Withholding of water   b) Pruning
c) Application of PGR's   d) Weed management

139. The most preferred carbon energy source in plant tissue culture is:
   a) Glucose b) Fructose
   c) Sucrose d) Maltose
140. 'Law of homologous series' was given by:
   a) Biffen b) Harlan
   c) Darwin d) Vavilov
141. An amino acid which is related to drought tolerance in plants is?
   a) Glycine b) Methionine
   c) Proline d) Cysteine
142. Pollen germination requires which of the following elements?
   a) B b) K
   c) Ca d) Si
143. The phenomenon in which germination of seeds is affected by light is known as:
   a) Photoperiodic b) Thermoperiodic
   c) Photoblastic d) Vernalized
144. In tissue culture, regeneration of and root occurs by the balance of:
   a) Auxin and ABA b) Auxin and cytokinin
   c) Cytokinin and ABA d) ABA and GA3
145. Which of the following phytohormones is used for the development of synthetic seeds?
   a) Auxin b) GA_3
   c) ABA d) Ethylene
146. Bioreactors are used for
   a) Production of ethanol
   b) Production of secondary metabolites
   c) Production of cell culture on large scale
   d) All of these
147. Which of the following fruit crops shows metaxenia?
   a) Apple b) Pear
   c) Peach d) Date palm
148. Phalsa seeds loose their viability under ordinary conditions after:
   a) 70-80 days b) 80-90 days
   c) 90-100 days d) 100-110 days

149. Iodine test is performed to judge the maturity index of:

a) Apple b) Guava

c) Mango d) Pineapple

150. Which plant has the highest longevity?

a) Walnut b) Pecan nut

c) Avocado d) Sweet chestnut

Answers Key

1.	b)	2.	a)	3.	a)	4.	c)	5.	d)	6.	d)	7.	c)
8.	b)	9.	d)	10.	b)	11.	c)	12.	c)	13.	c)	14.	c)
15.	a)	16.	a)	17.	a)	18.	b)	19.	b)	20.	b)	21.	c)
22.	b)	23.	c)	24.	c)	25.	c)	26.	b)	27.	a)	28.	c)
29.	b)	30.	b)	31.	b)	32.	d)	33.	a)	34.	d)	35.	a)
36.	b)	37.	a)	38.	b)	39.	b)	40.	b)	41.	a)	42.	d)
43.	b)	44.	c)	45.	d)	46.	d)	47.	b)	48.	b)	49.	d)
50.	b)	51.	b)	52.	b)	53.	c)	54.	c)	55.	b)	56.	b)
57.	d)	58.	b)	59.	a)	60.	b)	61.	b)	62.	a)	63.	c)
64.	a)	65.	c)	66.	a)	67.	b)	68.	c)	69.	c)	70.	b)
71.	d)	72.	a)	73.	b)	74.	b)	75.	a)	76.	c)	77.	b)
78.	d)	79.	d)	80.	b)	81.	c)	82.	c)	83.	b)	84.	c)
85.	c)	86.	c)	87.	a)	88.	c)	89.	b)	90.	b)	91.	a)
92.	d)	93.	d)	94.	c)	95.	b)	96.	b)	97.	b)	98.	c)
99.	b)	100.	d)	101.	b)	102.	d)	103.	b)	104.	c)	105.	a)
106.	a)	107.	c)	108.	d)	109.	d)	110.	b)	111.	b)	112.	d)
113.	a)	114.	c)	115.	d)	116.	c)	117.	c)	118.	a)	119.	b)
120.	b)	121.	d)	122.	b)	123.	c)	124.	a)	125.	c)	126.	a)
127.	d)	128.	b)	129.	b)	130.	b)	131.	c)	132.	d)	133.	d)
134.	d)	135.	c)	136.	c)	137.	b)	138.	d)	139.	c)	140.	d)
141.	c)	142.	a)	143.	c)	144.	b)	145.	c)	146.	d)	147.	d)
148.	c)	149.	a)	150.	d)								

9

ICAR – NET / ARS
Fruit Science Exam – 2014

1. Seedlessness in grapes is due to:
   a) Vegetative parthenocarpy b) Stimulative parthenocarpy
   c) Stenospermocarpy d) All of these
2. Bitterness in almond kernel is governed by:
   a) Single dominant gene b) Single recessive gene
   c) Mutagens d) Multiple genes
3. M-9 a dwarfing rootstock of apple has originated as a result of:
   a) Bud mutation b) Chance seedling selection
   c) Hybridization d) Clonal selection
4. The cross between *Euvitis* and *Muscadinia* is:
   a) Fertile b) Sterile
   c) Do not set fruits d) None of these
5. Which of the following is an apomictic rootstock species of apple?
   a) *Malus orientalis* b) *Malus sikkimensis*
   c) *Malus floribunda* d) *Malus sieversii*
6. Type of incompatibility reported in mango is:
   a) Sporophyte b) Gametophytic
   c) Both d) None of these
7. Kinnow mandarin is a first generation hybrid between:
   a) *Citrus nobilis* × *Citrus deliciosa*
   b) *Citrus nobilis* × *Citrus reticulata*
   c) *Citrus sinensis* × *Citrus deliciosa*
   d) *Citrus maxima* × *Citrus aurantium*
8. The ploidy level of cultivated strawberry (*Fragaria* x *ananassa*) is:
   a) Diploid b) Tetraploid
   c) Hexaploid d) Octaploid

9. Cultivated strawberry (*Fragaria* × *ananassa*) is a natural hybrid of:
   a) *Fragaria chiloensis* × *Fragaria vesca*
   b) *Fragaria chiloensis* × *Fragaria virginiana*
   c) *Fragaria virginiana* × *Fragaria vesca*
   d) *Fragaria virginiana* × *Fragaria chiloensis*
10. The ploidy level of Indian wild strawberry (*Fragaria vesca*) is:
    a) Diploid b) Tetraploid
    c) Hexaploid d) Octaploid
11. Grape fan leaf virus is spread by:
    a) Whitefly b) Aphid
    c) Bird d) Nematode
12. Grape rootstock 1613 is a hybrid of:
    a) *V. champinii* × *Othello* b) *V. solonis* × *Othello*
    c) *V. solonis* × *V. champinii* d) *Othello* × *V. solonis*
13. Novaria -a variety of banana has been developed using:
    a) Polyploidy b) *In vitro* mutagenesis
    c) Hybridization d) Embryo rescue
14. Novaria is an early flowering mutant of:
    a) Gross Michel b) Grand Nain
    c) Dwarf Cavendish d) Saba
15. Aonla cultivar, Amrit (NA-6) is a chance seedling selection of:
    a) Banarasi b) Hathijhool
    c) Chakaiya d) Balwant
16. Starkrimson (pear) is a bud mutant of:
    a) Beurre D'Anjou b) Bartlett
    c) Clapp's Favourite d) Doyenne de Cornice
17. A bud sport of Anab-e-Shahi is:
    a) Sharad Seedless b) Rao Sahebi
    c) Pandhari Sahebi d) Dilkhush
18. Star Ruby a deep red fleshed grapefruit has been developed by irradiation of seeds with thermal neutrons of which cultivar?
    a) Thompson b) Foster
    c) Hudson d) Ruby Red

19. Chakradhar is a thornless and seedless selection of:
    a) Acid lime
    b) Pummelo
    c) Grapefruit
    d) Mandarin
20. Nemaguard, is nematode resistant root stock of peach, is a hybrid between:
    a) *Prunus cerasifera* × *Prunus spinosa*
    b) *Prunus spinosa* × *Prunus davidiana*
    c) *Prunus cerasifera* × *Prunus persica*
    d) *Prunus persica* × *Prunus davidiana*
21. Red Elstar is a natural mutant of:
    a) Apple
    b) Cherry
    c) Grapefruit
    d) Nectarines
22. Embryo-sac sterility is found in:
    a) Cherry
    b) Sweet orange
    c) Grape
    d) Litchi
23. Which of the following strawberry varieties is known to be day neutral?
    a) Toiga
    b) Sweet Charlie
    c) Selva
    d) Pajaro
24. Foster pink is a bod sport of:
    a) Apple
    b) Peach
    c) Grapefruit
    d) Grape
25. Seedlessness in lemon (*Citrus limon*) is due to:
    a) Self-incompatibility
    b) Parthenocarpy
    c) Ovule sterility
    d) Embryo abortion
26. Hill banana belongs to which genomic group?
    a) AAA
    b) AAB
    c) ABB
    d) AB
27. Which of the following species of papaya is resistant to frost?
    a) *Vasconcella parviflora*
    b) *Vasconcella candamarcensis*
    c) *Vasconcella cauliflora*
    d) *Vasconcella microcarpa*
28. Ploidy level of Umran variety of ber (*Ziziphus mauritiana*) is:
    a) Diploid
    b) Tetraploid
    c) Hexaploid
    d) Octaploid

29. Which of the following is a …of red banana?

a) Namarai b) Mishri

c) Champa d) Saba

30. Which of the following is an example of polygamomonoecious plant:

a) Papaya b) Mango

c) Cashew nut d) Walnut

31. Which variety of grape sets fruit by stimulative parthenocarpy?

a) Perlette b) Thompson Seedless

c) Black Corinth d) Beauty Seedless

32. Bangalore Blue variety of grapes is a cross between:

a) *Vitis vulpina* × *Vitis labrusca*

b) *Vitis vinifera* × *Vitis rotundifolia*

c) *Vitis vinifera* × *Vitis labrusca*

d) *Vitis vinifera* × *Vitis munsoniana*

33. Seedless mango cv. Sindhu is a result of back cross between:

a) Neelum × Alphonso b) Ratna × Alphonso

c) Banganpalli × Alphonso d) Alphonso × Totapuri

34. Metaxenia is commonly observed in which fruit plant?

a) Apple b) Citrus

c) Date palm d) Loquat

35. Pollen sterility is a serious problem in:

a) Aonla b) Grape

c) Ber d) Bael

36. Which of the following is a dwarf mutant variety of papaya?

a) Pusa Dwarf d) Pusa Nanha

c) Pusa Majesty d) Pusa Giant

37. Protoandry is a problem in:

a) Grape b) Litchi

c) Walnut d) Date palm

38. Protogyny is found in which fruit crop?

a) Sapota b) Walnut

c) Passion fruit d) Coconut

39. Which variety of papaya produces only female plants:
    a) Pusa Delicious b) Sunrise Solo
    c) Taiwan d) Pink Flesh Sweet
40. Breba is a parthenocarpically produced:
    a) Spring crop in fig b) Summer crop in fig
    c) Rainy crop in fig d) None of these
41. Rosica variety of mango was developed through:
    a) Chance seedling selection b) Hybridization
    c) Mutation d) Ploidy manipulation
42. Baldwin, Gravenstin and Winesap cultivars of apple are in nature.
    a) Diploid b) Triploid
    c) Tetraploid d) Hexaploid
43. Popular male clone used in coffee breeding is:
    a) Hybrid-de-Timor b) Old chicks
    c) Taferikela d) Cauvery
44. Polyembryony has been reported in:
    a) Mango b) Jamun
    c) Citrus d) All of these
45. Genomic constitution of FHIA-1 (Gold Finger) is:
    a) AAAA b) ABBB
    c) AAAB d) ABB
46. Genomic classification of banana was given by:
    a) Tanaka and Swingle b) Simmonds and Shepherd
    c) Kostermans and Bompard d) Bhattacharya and Dutta
47. Mango hybridization work was first time started by:
    a) Majumdar and Mukherjee b) Iyer and Subramanyam
    c) Burns and Prayag d) Sharma and Singh
48. Ruby is a multiple hybrid of:
    a) Grapefruit b) Pomegranate
    c) Guava d) Grape
49. Who was an eminent breeder of grapes?
    a) H.P. Olmo b) T. Tanaka
    c) R.E. Litz d) Dennis Gonsalves

50. Which of the following is a teinturier variety of grape?
    a) Pusa Urvashi b) Pusa Navrang
    c) Punjab Purple d) Manjri Naveen
51. Apomictic seedlings are also known as:
    a) Nucellar seedling b) Zygotic seedling
    c) Polyembryonic seedling d) None of these
52. *Annona muricata* exhibits which type of dichogamy?
    a) Protoandry b) Protogyny
    c) Dichogamy d) Heterodichogamy
53. The type of incompatibility reported in loquat is:
    a) Sporophytic b) Gametophytic
    c) Genotypic d) All of these
54. Fruit size in mango is governed by:
    a) Single gene b) Poly genes
    c) Additive genes d) Complementary genes
55. Genome which is responsible for disease resistance in banana is:
    a) AA b) BB
    c) AB d) None of these
56. Mule tail/crazy top is a genetic disorder of:
    a) Almond b) Apricot
    c) Cherry d) Walnut
57. Which of the following fruit crop(s) is/are considered as manmade fruits?
    a) Kinnow b) Strawberry
    c) Atemoya d) All of these
58. The presence of anthocyanin strips in apple skin is controlled by:
    a) Single dominant gene b) Single recessive gene
    c) Additive genes d) Polygenes
59. Fuzziness in nectarine is controlled by:
    a) Single dominant gene b) Single recessive gene
    c) Additive genes d) Polygenes
60. Which of the following is a progenitor of apple?
    a) *Malus orientalis* b) *Malus baccata*
    c) *Malus sylvestris* d) *Malus sieversii*

61. Which of the following is a parthenocarpic variety of guava?
    a) Allahabad Surkha b) Allahabad Safeda
    c) Allahabad Round d) Arka Mridula
62. Which of the following is a seedless variety of grapefruit?
    a) Marsh b) Duncan
    c) Walters d) Foster
63. Most common type of hybrids in India are:
    a) D × T hybrid b) T × D hybrid
    c) T × T hybrid d) D × D hybrid
64. Citrange is a cross between:
    a) C. sinensis × Poncirus trifoliata
    b) C. paradisi × Poncirus trifoliata
    c) C. reticulata × Poncirus trifoliata
    d) C. aurantium × Poncirus trifoliata
65. Amritha is a hybrid variety of:
    a) Pomegranate b) Guava
    c) Grape d) Pineapple
66. Swarnroopa, a variety of litchi has been developed by:
    a) Chance seedling selection b) Clonal selection
    c) Mutation d) Hybridization
67. Which of the following varieties of litchi is resistant to fruit cracking?
    a) Shahi b) Rose Scented
    c) Swarnroopa d) Sabour Madhu
68. Mudkhed seedless is a bud mutant of:
    a) Kinnow mandarin b) Nagpur mandarin
    c) Khasi mandarin d) Coorg mandarin
69. Sloh, a peach and almond hybrid is:
    a) Self-sterile b) Self-fertile
    c) Cross sterile d) Cross fertile
70. Cayenne, a variety of pineapple, has .... chromosome number (2n=):
    a) 50 b) 75
    c) 100 d) 150

71. Duke cherry (*Prunus gaundini*) has ...... chromosome number:
    a) 2n=16 b) 2n=32
    c) 2n=48 d) 2n=64
72. Which of the following varieties is used as universal pollinizer for cherry?
    a) Governor Wood b) Bing
    c) Napoleon d) Compact Stella
73. Grapefruit (*Citrus paradisi*) is a natural hybrid between:
    a) Mandarin × Pummelo b) Sweet orange × Pummelo
    c) Sweet orange × Citron d) Sour orange × Citron
74. Sex form of Matua and Tomuri varieties of kiwifruit is:
    a) Staminate b) Pistillate
    c) Hermaphrodite d) Gynodioecious
75. Persimmon is a hexaploid with somatic chromosome number of:
    a) 30 b) 60
    c) 90 d) 120
76. Litchi and pistachio nut flowers are devoid of:
    a) Calyx b) Corolla
    c) Stamens d) Pistil
77. Which type of parthenocarpy has been reported in litchi?
    a) Vegetative b) Stimulative
    c) Stenospermocarpy d) All of these
78. Mango inflorescence contains ......type of flowers:
    a) Male and hermaphrodite b) Male and female
    c) Male and neutral d) Female and hermaphrodite
79. Allopolyploid has been reported in which cultivar of mango?
    a) Bappakai b) Vellaikolamban
    c) Goa d) Olur
80. J. H. Hale variety of peach is:
    a) Self-incompatible b) Self-sterile
    c) Self-unfruitful d) Self-fruitful
81. Which of the following is a self-sterile variety of pear?
    a) Magnese b) Flemish Beauty
    c) Bartlett d) Beurre Hardy

82. Pollination of fig flower by the *Blastophaga* wasp takes place through:
   a) Fascination
   b) Parthenogenesis
   c) Caprification
   d) Apomixis

83. Artificial hybridization in fruit crop in the world was first practiced by:
   a) T.A. Knight
   b) A.D. Candolle
   c) N.I. Vavilov
   d) Thomas Fairchild

84. Akshay is a superior clone of:
   a) Langra
   b) Dashehari
   c) Kesar
   d) Alphonso

85. Femaleness in papaya is governed by:
   a) $M_1 m^{RR}$
   b) $M_1 m$
   c) $M_2 m$
   d) mm

86. Improvement in papaya it done mainly through:
   a) Inbreeding and sibmatmg
   b) Polyploidy
   c) Hybridization
   d) Mutation

87. Cardinal grape is a cross between:
   a) Tokey × Ribier
   b) Ontario × Sultana
   c) Queen of Vineyard × Black Kishmish
   d) Himrod × Ribier

88. High Gate is a mutant of which commercial variety of banana?
   a) Gross Michel
   b) Dwarf Cavendish
   c) Grand Nain
   d) Poovan

89. European plum (*Prunus domestica*) has been originated as a natural hybrid between:
   a) *Prunus cerasifera* × *Prunus spinosa*
   b) *Prunus spinosa* × *Prunus davidiana*
   c) *Prunus cerasifera* × *Prunus persica*
   d) *Prunus persica* × *Prunus davidiana*

90. M.27 rootstock was evolved from ………….. cross:
   a) M.13 × M.7
   b) M.13 × M.9
   c) M.7 × M.13
   d) M.9 × M.13

91. Seeds are set without fertilization in:

a) Mango b) Avocado

c) Loquat d) Mangosteen

92. G-137, an improved variety of pomegranate is a selection from:

a) Muscat b) Ganesh

c) Alandi d) Jytoti

93. Flemish Beauty variety of pear is:

a) Self-sterile b) Self-fertile

c) Cross sterile d) Cross fertile

94. Which of the following species of papaya is resistant to virus?

a) *Vasconcella parviflora* b) *Vasconcella microcarpa*

c) *Vasconcella cauliflora* d) *Vasconcella pentagona*

95. Genome of reflexed stamen grapes is…………:

a) ShSf b) Sp

c) SoSp/SoSp d) So

96. Bodies Altafort, a tetraploid banana is a result of controlled crossing between:

a) Gross Michel × Psiang Lilin b) Gross Michel × Ney Poovan

c) Gross Michel × Kadli d) Gross Michel × Matti

97. CO-1 banana is a cross between:

a) Laden × Kadali b) Gross Michel × Kadli

c) Laden × Psiang Lilin d) Laden × Kadali × *Musa balbisiana*

98. The somatic chromosome number of mango is:

a) 2n=2x=40 b) 2n=4x=40

c) 2n=6x=40 d) 2n=8x=40

99 'Pusa Pitamber' a mango hybrid is cross between:

a) Amrapali × Sensation b) Dashehari × Sensation

c) Amrapali × Lal Sundari d) Amrapali ×Vanraj

100. Pusa Lalima is a cross between :

a) Amrapali × Sensation b) Amrapali × Lal Sundari

c) Dashehari × Sensation d) Dashehari × Lal Sundari

101. 1.5 g of IBA dissolved in 500 ml of 50% alcohol gives a concentration of:

a) 1500 ppm b) 3000 ppm

c) 4500 ppm d) 6000 ppm

102. High C:N ratio favours:

a) Vegetative growth b) Reproductive growth

c) Transpiration d) None of these

103. IBA is a most common growth regulator used for:

a) Improving fruit set b) Control of fruit drop

c) Rooting of cutting d) Ripening of fruits

104. A technique which has been found useful for producing disease free planting material of citrus is:

a) Stone grafting b) Softwood grating

c) Shoot tip grafting d) All of these

105. Zinc deficiency in grape causes:

a) Straggly cluster b) Bud killing

c) Berry drop d) Shriveled berry

106. Soil mulching helps in:

a) Reduction in evaporation b) Reduction in transpiration

c) Increase in evaporation d) Increase in transpiration

107. The most widely used chemical for killing the nematodes in nursery fields is:

a) Chloropicrin b) Formalin

c) Methyl bromide d) D-D mixture

108. Isogenic lines are developed through:

a) Back cross b) Pure line

c) Mutation d) Mass selection

109. IBPCR has now been renamed as:

a) World Genetic Resource Institute

b) Biodiversity International

c) Biodiversity International

d) Internal Plant Genetic Institute

110. Tatura trellis system of training was developed in:

a) USA b) Germany

c) New Zealand d) Australia

111. Which of the following is known as 'King of Arid Fruits'?

a) Aonla b) Ber

c) Bael d) Phalsa

112. Dwarf Cavendish variety of banana reaches maturity after .... days from shooting:

a) 90 ± 5 b) 100 ± 5

c) 110 ± 5 d) 120 ± 5

113. The storage temperature of banana is:

a) 10°C b) 13°C

c) 15°C d) 7°C

114. The undifferentiated mass of cells is known as:

a) Callus b) Explant

c) Media d) None of these

115. The progeny of a single plant obtained through asexual reproduction is known as:

a) Pure line b) Clone

c) Inbred d) Strain

116. Which of the following has highest calorific value?

a) Almond b) Banana

c) Date palm d) Cashew nut

117. High temperature during maturity period hampers fruit development and causes them to remain green in..........:

a) Sweet orange b) Kinnow mandarin

c) Khasi mandarin d) Blood Red Malta

118. The expanded form of CFB is:

a) Coloured Fiber Board b) Corrugated Filler Box

c) Corrugated Fiber Board d) Complete Filling Box

119. Kiwi fruit is commercially propagated by:

a) Air layering b) Hard wood cutting

c) Patch budding d) Seed

120. The 'Solen', a low domed form of training is particularly adopted for:

a) Spur bearing cultivars b) Non spur bearing cultivars

c) Tip bearing cultivars d) Lateral bearing cultivars

121. Indicator plant for exocortis in citrus is ....:

a) Acid lime b) Citron

c) Sweet orange d) Sweet line

122. The process of culturing rapidly growing shoot tips/root tips is known as:
 a) Meristem culture b) Cell culture
 c) Root culture d) Shoot culture
123. Which of the following is a transgenic variety of plum?
 a) Honey Dew b) Honey Crips
 c) Honey Sweet d) Honey pink
124. Which of the following is a naturally occurring auxin in plant?
 a) 2,4-D b) IAA
 c) IBA d) NAA
125. Which of the given plants is highly sensitive to moisture stress?
 a) Banana b) Grape
 c) Mango d) Citrus
126. Majority of the cultivated banana belongs to .... group:
 a) Cavendish b) Rasthali
 c) Monthan d) Hill banana
127. Tangor is an inter-specific cross between:
 a) Mandarin x Acid lime b) Mandarin x Grapefruit
 c) Mandarin x Sweet orange d) Mandarin x Sour orange
128. Central Institute of Arid Horticulture is located at:
 a) Jodhpur b) Udaipur
 c) Bikaner d) Jaipur
129. Predominant organic acid present in grape is:
 a) Citric b) Tartaric
 c) Oxalic d) None of these
130. Which of the following is the richest source of vitamin C?
 a) Aonla b) Guava
 c) Citrus d) Barbados cherry
131. In Navel orange, navel is present 'at the:
 a) Thalamic end b) Stylar end
 c) Micropylar end d) None of these
132. Hot water treatment of hard seed to make them permeable for water and air is known as:
 a) Stratification b) Scarification
 c) Seed priming d) None of these

133. The sequential cutting of banana pseudo-stem after harvesting is known as:
   a) Propping b) Desuckering
   c) Mattocking d) Plucking
134. In north India, pruning in grape is done in the month of:
   a) October b) January
   c) March d) June
135. Disease free plants in micro-propagation can be obtained through:
   a) Anther culture b) Ovule culture
   c) Meristem culture d) Embry rescue
136. Which fruit crop has the highest chromosome number?
   a) Mulberry b) Persimmon
   c) Cashew nut d) Passion fruit
137. Which one of the following is a non-climacteric fruit?
   a) Banana b) Sapota
   c) Pineapple d) Kiwifruit
138. The deficiency of which of the following causes lime induced chlorosis?
   a) Zn b) Mg
   c) Fe d) S
139. Roopa is an improved variety of:
   a) Litchi b) Walnut
   c) Grape d) Pomegranate
140. Pineapple is a/an:
   a) Annual herb b) Perennial herb
   c) Annual climber d) Perennial climber
141. The term 'Chicken Tongue' is associated with:
   a) Grape b) Papaya
   c) Litchi d) Cherry
142. Which Indian state has the highest area under protected cultivation in India?
   a) Andhra Pradesh b) Himachal Pradesh
   c) Maharashtra d) Punjab
143. The best irrigation system for protected cultivation is:
   a) Sprinkler system b) Drip system
   c) Basin system d) Flood system

144. Which of the following crop is highly susceptible to salinity?

a) Mango b) Citrus

c) Grape d) Date palm

145. Which cladding material provides the maximum light transmittance in greenhouse?

a) Glass b) Polythene

c) Polyvinyl chloride d) Acrylic

146. Drip irrigation in India was introduced from:

a) Germany b) California

c) Israel d) Iran

147. Application of fertilizers along with irrigation water is known as:

a) Fumigation b) Fertigation

c) Fertilization d) None of these

148. Which of the following plants is highly susceptible to water logging?

a) Papaya b) Guava

c) Banana d) Grape

149. Agri-export zone identified for walnut is in:

a) Jammu & Kashmir b) Himachal Pradesh

c) Punjab d) Tripura

150. Nendran banana belongs to which group?

a) AAA b) AAB

c) ABB d) AB

Answers Key

1.	(c)	2.	(b)	3.	(b)	4.	(b)	5.	(b)	6.	(a)	7.	(a)
8.	(d)	9.	(b)	10.	(a)	11.	(d)	12.	(b)	13.	(b)	14.	(b)
15.	(c)	16.	(c)	17.	(d)	18.	(c)	19.	(a)	20.	(d)	21.	(a)
22.	(b)	23.	(c)	24.	(c)	25.	(a)	26.	(b)	27.	(b)	28.	(b)
29.	(b)	30.	(c)	31.	(c)	32.	(c)	33.	(b)	34.	(c)	35.	(c)
36.	(b)	37.	(c)	38.	(a)	39.	(b)	40.	(a)	41.	(c)	42.	(b)
43.	(c)	44.	(d)	45.	(c)	46.	(b)	47.	(c)	48.	(b)	49.	(a)
50.	(b)	51.	(a)	52.	(a)	53.	(b)	54.	(b)	55.	(b)	56.	(a)
57.	(d)	58.	(a)	59.	(b)	60.	(d)	61.	(c)	62.	(a)	63.	(a)
64.	(a)	65.	(d)	66.	(a)	67.	(c)	68.	(b)	69.	(b)	70.	(b)
71.	(b)	72.	(d)	73.	(b)	74.	(a)	75.	(b)	76.	(b)	77.	(b)
78.	(a)	79.	(b)	80.	(b)	81.	(a)	82.	(c)	83.	(a)	84.	(b)
85.	(d)	86.	(a)	87.	(a)	88.	(a)	89.	(a)	90.	(d)	91.	(d)
92.	(b)	93.	(b)	94.	(c)	95.	(c)	96.	(a)	97.	(d)	98.	(a)
99.	(c)	100.	(c)	101.	(b)	102.	(b)	103.	(c)	104.	(c)	105.	(a)
106.	(a)	107.	(d)	108.	(a)	109.	(c)	110.	(d)	111.	(b)	112.	(b)
113.	(b)	114.	(a)	115.	(b)	116.	(a)	117.	(b)	118.	(c)	119.	(b)
120.	(c)	121.	(b)	122.	(a)	123.	(c)	124.	(b)	125.	(a)	126.	(a)
127.	(c)	128.	(c)	129.	(b)	130.	(d)	131.	(b)	132.	(b)	133.	(c)
134.	(b)	135.	(c)	136.	(a)	137.	(c)	138.	(c)	139.	(b)	140.	(b)
141.	(c)	142.	(c)	143.	(b)	144.	(b)	145.	(a)	146.	(c)	147.	(b)
148.	(a)	149.	(a)	150.	(b)								

10

ICAR – NET / ARS Fruit Science Exam – 2015

1. Shot berry in Perlette grape can be overcome by spray of ethrel @:

   a) 15 ppm b) 20 ppm

   c) 25 ppm d) 30 ppm

2. Which of the following plant growth regulator is used for somatic embryogenesis in mango?

   a) Kinetin b) 2, 4-D

   c) GA3 d) ABA

3. Which of the following fruit crops has the highest chilling requirement?

   a) Cherry b) Apple

   c) Walnut d) Apricot

4. In vitro mutagenesis is useful in:

   a) Grape b) Banana

   c) Pineapple d) Citrus

5. Agro biodiversity hotspots in India are:

   a) 15 b) 18

   c) 22 d) 25

6. Geographical Indication (GI) tag is valid for:

   a) 8 year b) 10 year

   c) 15 year d) 18 year

7. Under UPOV system, duration of protection for tree and vine crops are:

   a) 15 year b) 18 year

   c) 20 year d) 25 year

8. Synersis of jelly is caused due to:

   a) Excess of acid b) Overcooking

   c) Excess of pectin d) Lack of pectin

9. Criteria of essentiality of mineral elements was given by:
   a) Arnon and Stout b) Liebig
   c) Brown et al. d) Munch
10. The movement of auxin in stem (aerial) parts of plants is:
   a) Acropetal b) Basipetal
   c) Non-directional d) None of these
11. The words 'in vitro' and 'in vivo' are:
   a) Latin b) English
   c) Greek d) Persian
12. Femaleness in papaya is governed by:
   a) M,mRR b) M,m
   c) M2m d) mm
13. The cross between hermaphrodite x hermaphrodites in papaya gives progeny of:
   a) 1:1 hermaphrodite and female b) 2:1 female and hermaphrodite
   c) 2:1 hermaphrodite and female d) 3:1 hermaphrodite and female
14. Agri Export Zone identified for Pineapple is:
   a) Tripura b) Kerala
   c) Assam d) Meghalaya
15. Agri Export Zone identified for Pomegranate is:
   a) Maharashtra b) Gujrat
   c) Madhya Pradesh d) Andhra Pradesh
16. Agri Export Zone identified for mango and grapes is:
   a) Maharashtra b) Andhra Pradesh
   c) Gujarat d) Tamil Nadu
17. Lincoln canopy method of training was developed in:
   a) Australia b) Italy
   c) Germany d) New Zealand
18. Which of the following is known as 'Buddha's hand fruit'?
   a) Banana b) Citrus
   c) Jackfruit d) Durian

19. Dormancy caused due to photogenic factors outside the affected - structure is known as:
    a) Ecodormancy b) Endodormancy
    c) Paradormancy d) AH the above
20. Which of the following is the highest salt tolerant fruit crop?
    a) Datepalm b) Ber
    c) Aonla d) Guava
21. A herbaceous perennial, monocotyledon, monoecious, mesophytes fruit crop is:
    a) Pineapple b) Strawberry
    c) Banana d) Papaya
22. Coat protein mediated resistance has been developed in:
    a) Papaya b) Banana
    c) Citrus d) All the above
23. Which of the following is an ornithophilous fruit crop?
    a) Pineapple b) Banana
    c) Strawberry d) Oil palm
24. Which of the following is known as 'Ripening Hormone1?
    a) Auxin b) Cytokinin
    c) Gibberellins d) Ethylene
25. Mangla is a variety of:
    a) Arecanut b) Coffee
    c) Tea d) Coconut
26. Which of the following rootstock is resistant to fire blight of apple?
    a) M.9 b) EMLA.9
    c) MM.106 d) Ottawa
27. The total area under drip irrigation in India is:
    a) 15.6 lac ha b) 18.9 lac ha
    c) 21.3 lac ha d) 24.6 lac ha
28. Which of the following is used for berry elongation in grapes?
    a) NAA b) Kinetin
    c) Dormex d) GA_3

29. Fruit thinning hormone is:
    a) NAA b) Ethrel
    c) DNOC d) Gibberellins
30. Which of the following is used for vigour control in apple?
    a) SADH b) 2 4 D
    c) ABA d) MH
31. Highest production and productivity of apple in India is in:
    a) HP b) J&K
    c) Uttarakhand d) Punjab
32. Optimum pH for orchard planting is:
    a) 5.5-6.5 b) 6.5-7.5
    c) 7.5-8.5 d) 8.5-9.5
33. Cybrid is a/an:
    a) Apomictic hybrids b) Somatic hybrids
    c) Cytoplasmic hybrids d) All the above
34. The viability of papaya seed is:
    a) 30 days b) 45 days
    c) 75 days d) 90 days
35. Which of the following is a self-incompatible mango variety?
    a) Amrapali b) Langra
    c) Alphonso d) Neelum
36. Mango is commercially propagated by:
    a) Wedge grafting b) Epicotyl grafting
    c) Tongue grafting d) Saddle grafting
37. Pomegranate is commercially propagated by:
    a) Air layering b) Hard wood cutting
    c) Wedge grafting d) Semi hard wood cutting
38. Litchi has originated from:
    a) Northern China b) Southern China
    c) Central China d) Eastern China
39. Cocoa is native of:
    a) North America b) South America
    c) Malaya archipelago d) West Africa

40. Pusa Majesty and Pusa Delicious varieties of papaya are:
    a) Dioecious b) Monoecious
    c) Gynodioecious d) Hermaphrodite
41. Dwarf Cavendish varieties of banana reach maturity after days from shooting:
    a) 90 ± 5 b) 100 ± 5
    c) 110 ± 5 d) 120 ± 5
42. Protogynous dichogamy is found in:
    a) Coconut b) Walnut
    c) Annona d) Chestnut
43. Which of the following fruit crops has no named varieties?
    a) Annona b) Avocado
    c) Loquat d) Mangosteen
44. Cherry seeds which can tolerate loss of moisture and their longevity can be enhanced by preserving them at low temperature are known as:
    a) Orthodox seed b) Recalcitrant seed
    c) Cryogenic seed d) All the above
45. Which of the following is an abiotic stress?
    a) Cold b) Virus
    c) Bacteria d) Fungi
46. J. H. Hale variety of peach is:
    a) Self-incompatible b) Self-sterile
    c) Self-unfruitful d) Self-fruitful
47. Gibberellins enhance the sugar level of fruits by causing:
    a) Decreased amylase activity b) Increased amylase activity
    c) Increasing acid content d) None of the above
48. A citrus variety most susceptible to granulation is:
    a) Kinnow b) Jaffa
    c) Nagpur Mandarin d) Mosambi
49. Which of the following is/ are low chilling peaches?
    a) Sharbati b) Shan e Punjab
    c) Florda Sun d) All of these

50. Fuji is a variety of:
    a) Apple b) Apricot
    c) Almond d) Cherry

51. Which of the following apple varieties is resistant to apple scab:
    a) Akbar b) Firdaus
    c) Lal Ambri d) Sunehari

52. Which of the following apple varieties is earliest in ripening:
    a) Fuji b) Red June
    c) Royal Delicious d) Golden Delicious

53. Pixy is a dwarfing rootstock of:
    a) Peach b) Cherry
    c) Plum d) Apricot

54. Very high temperature in fig causes:
    a) Discoloration of flesh b) Premature ripening
    c) Small fruits d) None of the above

55. Which of the following is a terminal bearer fruit crop?
    a) Mango b) Citrus
    c) Guava d) Peach

56. Single sigmoid growth curve is found in:
    a) Apple b) Peach
    c) Seedless banana d) Fig

57. Triple sigmoid growth curve is found in:
    a) Cherry b) Apricot
    c) Passion fruit d) Kiwifruit

58. Kniffin system of training is originally a:
    a) Two cane system b) Four cane system
    c) Six cane system d) None of the above

59. Training system followed in passion fruit is:
    a) T-bar b) Two arm kniffin
    c) Matted row d) Four arm kniffin

60. Kiwi fruit is trained through:
    a) T-bar b) Two arm kniffin
    c) Matted row d) Four arm kniffin

61. Which of the following fruit crops does not require pruning?
    a) Sapota b) Phalsa
    c) Guava d) Apple
62. Which of the following fruit crops has large flowering time?
    a) Avocado b) Custard apple
    c) Jamun d) Mangosteen
63. Which of the following is a temporary preservation method?
    a) Pasteurization b) Sterilization
    c) Canning d) Fermentation
64. Blanching of fruits is done at:
    a) 80°C b) 90°C
    c) 100°C d) 120°C
65. Canning is related to:
    a) Preservation b) Disease management
    c) Physiological disorder d) Irrigation
66. CF (Grand Ferrade) series of rootstocks are related to:
    a) Apricot b) Cherry
    c) Peach d) Plum
67. Which of the following is an abiotic stress:
    a) Fungus b) Virus
    c) Bacteria d) High temperature
68. In citrus classification, which of the following is known as lumper?
    a) Tanaka b) Swingle
    c) Hodgson d) Bhattacharya and Dutta
69. Training method followed in ber:
    a) Open vase b) Central leader
    c) Modified central leader d) Multi stem training
70. Monogenic resistant to apple scab is:
    a) Vf b) Va
    c) Vm d) Vr
71. Water saving in banana due to drip irrigation is:
    a) 20-30% b) 30-40%
    c) 40-50% d) 50-60%

72. Storage temperature of banana:
    a) 10°C b) 13°C
    c) 15°C d) 7°C

73. Controlled atmospheric storage is mostly used in:
    a) Apple b) Grape
    c) Citrus d) Mango

74. O, the concentration in control atmospheric storage of apple is:
    a) <2% b) <5%
    c) <7% d) <10%

75. Virus free plants are produced from:
    a) Meristem culture b) Embryo culture
    c) Anther culture d) Ovule culture

76. Pusa Shrestha mango is a hybrid of:
    a) Dashehari × Sensation b) Amrapali × Lal Sundari
    c) Amrapali × Sensation d) Amrapali × Vanraj

77. Which of the following is a spur pruning grape variety?
    a) Perlette b) Anab-e-Shahi
    c) Thompson Seedless d) Bhokri

78. Which of the following is a cane pruning grape variety?
    a) Perlette b) Anab-e-Shahi.
    c) Delight d) Bangalore Blue

79. HCN in grape is used to:
    a) Hasten early bud break b) Berry elongation
    c) Increase fruitfulness d) Control vegetative growth

80. Mridula pomegranate is a hybrid of:
    a) Ganesh × Gulshared
    b) Ganesh × "Nana
    c) Open pollinated F, of Ganesh × Gulshared
    d) Open pollinated F, of Ganesh × Nana

81. Which of the following is a polyembryonic fruit crop?
    a) Pomelo b) Karonda
    c) Citron d) Citrus maxima

82. Number of plants accommodated in a hectare of meadow orcharding are:
    a) 2000 b) 5000
    c) 7000 d) 10000
83. Fruit crop which shares the highest production in India is:
    a) Mango b) Citrus
    c) Papaya d) Banana
84. Carambola (*Averrhoa carambola*) belongs to family:
    a) Moraceae b) Rosaceae
    c) Oxalidaceae d) Rutaceae
85. Acidic nature of star fruit is due to:
    a) Propanoic b) Malic acid
    c) Oxalic acid d) Linoleic acid
86. The native place of jackfruit is:
    a) Bangladesh b) Indonesia
    c) Philippines d) India
87. Chromosome number (2n=) of her is:
    a) 12 b) 24
    c) 36 d) 48
88. The fruit crop which is subtropical in nature but can be grown successfully in temperate region is:
    a) Peach b) Plum
    c) Strawberry d) Persimmon
89. Which of the following is a non-climacteric fruit crop?
    a) Mango b) Apple
    c) Kiwifruit d) Litchi
90. Guava bears mostly on:
    a) One year old shoot b) Current season growth
    c) Spurs d) Very old shoot
91. 'Roopa' is an improved variety of:
    a) Litchi b) Walnut
    c) Jackfruit d) Peach
92. Which of the following is a seedless variety of sweet orange?
    a) Blood Red Malta b) Hamlin
    c) Washington Navel d) Mosambi

93. Unreduced or restitution gametes are formed in:

a) Mango c) Citrus

b) Banana d) Guava

94. Mango inflorescence contains which type of flowers:

a) Male and hermaphrodite b) Male and female

c) Male and neutral d) Female and hermaphrodite

95. Aneuploidy has been successfully used to improve:

a) Guava b) Papaya

c) Banana d) Grape

96. Kew is a variety of:

a) Litchi b) Avocado

c) Pineapple d) Loquat

97. 'Bodies Altafort' variety of banana is:

a) Diploid b) Triploid

c) Triploid d) Hexaploid

98. Cricket Ball x Oval are the parents of which of the following sapota varieties:

a) CO-1 b) CO-2

c) CO-3 d) PKM-l

99. Mango malformation can be overcome by application of:

a) NAA @100 ppm b) NAA @200 ppm

c) 2,4-D @200 ppm d) Paclobutrazol @200 ppm

100. Double row system of planting is followed in:

a) Banana b) Pineapple

c) Papaya d) Apple

101. Which of the following elements is highly mobile in plants?

a) N b) Zn

c) Ca d) Fe

102. Pollen sterility is a serious problem in:

a) Ber b) Aonla

c) Grape d) Pomegranate

103. Which type of fig bears 'Breba crop'?

a) San Pedro fig b) Common fig

c) Smyrna fig d) Capri fig

104. Tatura trellis training system is commonly used in:

a) Apple b) Plum

c) Peach d) Pear

105. Protoplast fusion/parasexual hybridization is used to produce:

a) Cybrids b) Hybrids

c) Soma clones d) None of these

106. Embryo rescue technique can be used for production of seedless ....:

a) Apples b) Litchi

c) Grapes d) Pomegranate

107. Which of the following has non-endospermic seed?

a) Bael b) Pear

c) Apple d) Ber

108. Cauliflory is observed in:

a) Jackfruit b) Carambola

c) Durian d) All of these

109. Which of the following is a thermoacoustic vitamin?

a) Vitamin A b) Vitamin B

c) Vitamin C d) Vitamin D

110. A plant or plant part composed of genetically different layer is known as:

a) Chimera b) Mutant

c) Hybrid d) Variant

111. Mist chamber is used for:

a) Air layering b) Rooting of cutting

c) Budding d) Grafting

112. The thickness of polythene sheet used in plant propagation is:

a) 100 gauge b) 200 gauge

c) 300 gauge d) 400 gauge

113. Which of the following plant growth regulator is used to induce rooting in cutting?

a) IAA b) NAA

c) IBA d) 2-4 D

114. Which of the following is a storage disorder of apple?

a)	Cork spot	b)	Water core
c)	Bitter pit	d)	Jonathan spot

115. Macadamia nut (*Macadamia ternifolia*) was introduced in India from:

a)	Australia	b)	USA
c)	Philippines	d)	Japan

116. Top working in temperate fruit plants is best done by the technique:

a)	Tongue grafting	b)	Cleft grafting
c)	Veneer grafting	d)	Chip budding

117. Astringency in persimmon is lost due to:

a) Conversion of tannins to sugar

b) Polymerization of tannins

c) Breakdown of mannitol to sugars

d) Breakdown of polysaccharide to sugars

118. Ethylene is responsible for the development of in fruits.

a)	Aroma	b)	Color of flesh
c)	Texture	d)	Shape

119. In litchi. heavy reiterative pruning is followed usually up to limbs at a height of:

a)	I-2m	b)	2-3m
c)	4-5m	d)	6-8m

120. Which of the following is a polygamous fruit crop?

a)	Guava	b)	Mango
c)	Papaya	d)	Mandarins

121. Elimination of nucleus of one species from a normal heterokaryon occurs in:

a)	Cybrids	b)	Haploids
c)	Hybrids	d)	Mutants

122. Which of the following methods is used for preservation of germplasm of fruit crops?

a)	*In-situ* conservation	b)	*Ex-situ* conservation
c)	On farm conservation	d)	None of these

123. Which of the following is a major constraint in mango cultivation in tropical semi-arid regions?

a) Frost | b) Cold
c) Heat | d) Draught

124. Double sigmoid growth curve is observed in:

a) Apple | b) Pear
c) Almond | d) Peach

125. The characteristic of an ideal glaring material is that it should:

a) Transmit the visible portion of sunlight
b) Absorb the ultra violet portion and convert some of it into visible light
c) Reflect or absorb infra-red radiation
d) All of these

126. Which of the following glaring materials provides maximum light transmittance?

a) Acrylic | b) Polycarbonate
c) Polyethylene | d) glass

127. The thickness of I'V stabilized plastic film used for the construction of low-cost wooden framed plastic film greenhouse is:

a) 100 micron | b) 200 micron
c) 300 micron | d) 400 micron

128. Which of the following is an immobile element in plants?

a) Zn | b) Ca
c) Mn | d) Fe

129. How much amount of GA3 is required for preparation of 5 ppm in 5 litre water?

a) 2.5 mg | b) 25 mg
c) 50 mg | d) 250 mg

130. Dwarfness in papaya is controlled by:

a) Recessive gene | b) Additive gone
c) Dominant gene | d) Duplicate gene

131. CAM cycle is found in which fruit crop?

a) Sapota | b) Strawberry
c) Pineapple | d) Papaya

132. Crop regulation (bahar treatment) in citrus and pomegranate is achieved by:
   - a) With holding of irrigation water
   - b) Root and shoot pruning
   - c) Deblossoming by application of NAA
   - d) All of these

133. Fruit cracking in pomegranate occurs due to the deficiency of:
   - a) Zn
   - b) Mg
   - c) Mn
   - d) B

134. Inflorescence type of cashew nut is:
   - a) Polygamomonoecious
   - b) Andromonoecious
   - c) Gynomonoecious
   - d) Trimonoecious

135. Among the following, pre-cooling is related to:
   - a) Preservation
   - b) Freezing
   - c) Cut flowers
   - d) None of these

136. Which of the following glycoside is present in sour orange:
   - a) Aurantemanm
   - b) Narmgine
   - c) Hesperian
   - d) Annontn

137. Cider is a fermented product of:
   - a) Apple
   - b) Pear
   - c) Datepalm
   - d) Cashew apple

138. Micro-propagation was first standardized for:
   - a) Almond
   - b) Banana
   - c) Grape
   - d) Strawberry

139. High cytokine in/auxin ratio in callus favours:
   - a) Shoot formation
   - b) Root formation
   - c) Flowering
   - d) Organogenesis

140. Nectarines are:
   - a) Smooth skinned peaches
   - b) Dried almonds
   - c) Smooth skinned plums
   - d) Smooth skinned apricots

141. Seedlessness in grape is controlled by:
   - a) Dominant gene
   - b) Recessive gene
   - c) Complementary gene
   - d) Duplicate gene

142. Which of the following chemicals is used for polyploidy breeding?

a) Colchicine b) MH

c) NAA d) SADH

143. Which of the following is the chief cause of spreading decline in citrus?

a) Citrus nematode b) Burrowing nematode

c) Fungal diseases d) Nutritional deficiency

144. Neelum/Alphonso cross in mango results in the development of:

a) Ratna b) Sindhu

c) Arka Puneet d) None of these

145. Maximum heating is obtained with which mulch material?

a) Black mulch b) White mulch

c) Straw d) Saw dust

146. Which chemical acts as an aid for mechanical harvesting?

a) GA, b) IAA

c) ABA d) Ethylene

147. Seededness in banana can be reduced by spraying:

a) GA, b) Ethephon

c) 2,4-D d) Kinein

148. *Ziziphus nummularia* is known to impart dwarfness in ber. Dwarfness is caused due to:

a) Graft incompatibility b) Poor nutrient uptake

c) Poor root formation d) Virus infection

149. The ancestor of *Ananas comosus* is:

a) *Ananas bractus* b) *Ananas microstachys*

c) *Ananas erectifolius* d) *Ananas ananasoides*

150. Predominant organic acid present in grape is:

a) Citric b) Tartaric

c) Oxalic d) None of these

Answers Key

1.	(c)	2.	(b)	3.	(a)	4.	(b)	5.	(c)	6.	(b)	7.	(d)
8.	(a)	9.	(a)	10.	(b)	11.	(a)	12.	(d)	13.	(c)	14.	(a)
15.	(a)	16.	(b)	17.	(d)	18.	(b)	19.	(c)	20.	(a)	21.	(b)
22.	(d)	23.	(a)	24.	(d)	25.	(a)	26.	(d)	27.	(b)	28.	(d)
29.	(a)	30.	(a)	31.	(b)	32.	(b)	33.	(c)	34.	(d)	35.	(b)
36.	(a)	37.	(b)	38.	(b)	39.	(b)	40.	(c)	41.	(a)	42.	(c)
43.	(d)	44.	(a)	45.	(a)	46.	(b)	47.	(b)	48.	(d)	49.	(d)
50.	(a)	51.	(b)	52.	(b)	53.	(c)	54.	(b)	55.	(a)	56.	(a)
57.	(d)	58.	(b)	59.	(b)	60.	(a)	61.	(a)	62.	(b)	63.	(a)
64.	(c)	65.	(a)	66.	(d)	67.	(d)	68.	(b)	69.	(c)	70.	(a)
71.	(c)	72.	(b)	73.	(a)	74.	(b)	75.	(a)	76.	(c)	77.	(a)
78.	(b)	79.	(a)	80.	(c)	81.	(a)	82.	(b)	83.	(d)	84.	(c)
85.	(c)	86.	(d)	87.	(d)	88.	(d)	89.	(d)	90.	(b)	91.	(b)
92.	(c)	93.	(b)	94.	(a)	95.	(a)	96.	(c)	97.	(c)	98.	(a)
99.	(b)	100.	(b)	101.	(a)	102.	(a)	103.	(a)	104.	(c)	105.	(a)
106.	(c)	107.	(d)	108.	(a)	109.	(c)	110.	(a)	111.	(b)	112.	(b)
113.	(c)	114.	(c)	115.	(a)	116.	(b)	117.	(a)	118.	(a)	119.	(a)
120.	(c)	121.	(a)	122.	(b)	123.	(a)	124.	(d)	125.	(d)	126.	(d)
127.	(b)	128.	(b)	129.	(b)	130.	(a)	131.	(c)	132.	(d)	133.	(d)
134.	(a)	135.	(a)	136.	(a)	137.	(a)	138.	(d)	139.	(a)	140.	(a)
141.	(b)	142.	(a)	143.	(b)	144.	(a)	145.	(a)	146.	(c)	147.	(c)
148.	(a)	149.	(b)	150.	(b)								

11

ICAR – NET Fruit Science Exam – 2016

1. Which rare species of *Mangifera* is endemic to the Philippines
   - a) *Mangifera merrillii*
   - b) *Mangifera minutifolia*
   - c) *Mangifera laurina*
   - d) *Mangifera paludosa*
2. Which of the following crops is largely exported from India?
   - a) Walnut
   - b) Mango
   - c) Grape
   - d) Banana
3. Lowest number of AEZs belongs which state?
   - a) Maharashtra
   - b) Karnataka
   - c) West Bengal
   - d) Andhra Pradesh
4. The edible part of walnut is:
   - a) Endosperm
   - b) Lobed cotyledons
   - c) Pericarp
   - d) Epicarp
5. Which of the following countries is the greatest producer of almond in the world?
   - a) USA
   - b) China
   - c) Spain
   - d) Afghanistan
6. The sweetest sugar in fruit is:
   - a) Sucrose
   - b) Glucose
   - c) Fructose
   - d) Lactose
7. NBA (National Biodiversity Authority) is headquartered in:
   - a) Chennai
   - b) New Delhi
   - c) Bangalore
   - d) Hyderabad
8. Arctic apple (non-browning apple) is developed through:
   - a) Gene silencing
   - b) Protoplast fusion
   - c) *In vitro* mutagenesis
   - d) Agrobacterium mediated transformation

9. The agency responsible for introduction and maintenance of germplasm of fruit crops is:
   a) NBPGR b) ICAR
   c) BSI d) FRI
10. Which of the following countries has the largest area under greenhouse:
   a) China b) Japan
   c) USA d) Spain
11. Mauritius, a popular variety of pineapple is classified as:
   a) Early season b) Mid season
   c) Late season d) Very late season
12. Triploid growth curve is observed in:
   a) Kiwifruit b) Pecan nut
   c) Cherry d) Carambola
13. Alphonso variety of mango is mostly grown in:
   a) Northern India b) Southern India
   c) Eastern India d) Western India
14. Guava bears mostly on:
   a) Current season growth b) One year old shoot
   c) Spur d) Very old shoot
15. Lalit variety of guava is developed through:
   a) Selection b) Hybridization
   c) Mutation d) Transgenic
16. The inception of ripening in grape is termed as:
   a) Breakeven point b) Verasion
   c) Colour break d) Turning stage
17. is known as the 'Food of God'
   a) Tea b) Coffee
   c) Cocoa d) Cashew
18. Which of the following mango varieties is resistant to mango malformation?
   a) Ellaichi b) Bombay green
   c) Dashehari d) Langra
19. Sweet orange is generally trained on:
   a) Single stem b) Multiple stem
   c) Two branches d) None of these

20. Pyronia is a cross between:
    a) Pear × Quince
    b) Apple × Pear
    c) Pear × Loquat
    d) Apple × Quince
21. Which of the following is an ethylene absorbent?
    a) KMnO4
    b) KNO3
    c) K2SO4
    d) KCl
22. Inarching in mango in India was suggested by:
    a) H. Lynch
    b) R.N. Singh
    c) S.K. Mukherjee
    d) R.M. Pandey
23. Based on the climatic requirements, carambola is classified as a:
    a) Tropical fruit
    b) Sub-tropical fruit
    c) Temperate fruit
    d) Sub-temperate fruit
24. Bitter pit in apple is caused due to the deficiency of:
    a) Mn
    b) Ca
    c) B
    d) Zn
25. Markers used for determining genetic variation in plants is:
    a) Morphological markers
    b) Biochemical markers
    c) Molecular markers
    d) All of these
26. Alcohol content in apple cider is:
    a) 2-4%
    b) 4-6%
    c) 6-8%
    d) 8-10%
27. Sugar used as preservative concentration of:
    a) >45%
    b) >55%
    c) >65%
    d) >75%
28. The term 'Cold Sterilization' is also known as:
    a) Ultra filtration
    b) Irradiation
    c) Refrigeration
    d) Freezing
29. Which of the following is known as 'self- peeling banana'?
    a) *Musa laterita*
    b) *Musa ingens*
    c) *Musa basjoo*
    d) *Musa velutina*
30. Which of the following is a polygenic source of resistant to apple scab?
    a) Antonovka
    b) *Malus floribunda*
    c) *Malus micromalus*
    d) *Malus pumila*

31. An example of monoaxial fruit plant is:
    a) Pomegranate b) Cherry
    c) Phalsa d) Papaya
32. The stage of inflorescence emergence in pineapple is known as:
    a) Sorose b) Piping
    c) Rosette d) Red heart

    Richmond-Lang effect describes one of the basic properties of which plant growth regulator:
    a) Jasmonic acid b) Cytokinin
    c) Ethylene d) Abscisic acid
34. Cauliflory is observed in:
    a) Jackfruit b) Peach
    c) Loquat d) All of these
35. Yellow spot, a physiological disorder of citrus occurs due to the deficiency of:
    a) Zn b) B
    c) Mo d) Ca
36. Rainbow and Sunup are the transgenic varieties of:
    a) Plum b) Papaya
    c) Banana d) Apple
37. The most noticeable impact of climate change on mango trees in India is:
    a) Heavy fruit drop b) Erratic flowering
    c) Incidence of mango malformation d) Bark splitting
38. Flowering in aonla occurs on which type of shoots?
    a) Determinate b) Indeterminate
    c) Apical d) All of the above
39. Which of the following is the richest source of Omega-3 fatty avcids?
    a) Grape b) Olive
    c) Walnut d) Almond
40. An ideal fruit crop for edible vaccine through genetic transformation is:
    a) Red banana b) Pineapple
    c) Papaya d) Tree tomato

41. The major problem in phalsa cultivation is considered to be:
    a) Non-synchronized ripening b) Susceptibility to foliar diseases
    c) Self-incompatibility d) Male sterility
42. Which of the following variety of mandarin is commercially grown in northern India?
    a) Kinnow mandarin b) Coorg mandarin
    c) Nagpur mandarin d) Khasi mandarin
43. Which fruit is used for preparation of good quality jelly?
    a) Mango b) Jamun
    c) Aonla d) Guava
44. Flavouring compound in over ripe banana is:
    a) Amyl acetate b) Isopentanol
    c) Flavonoids d) Hexanal
45. Salt concentration in pickle is measured by:
    a) Refractometer b) Salt meter
    c) Solometer d) pH meter
46. Browning in apple occurs due to the activity of:
    s) Hydrogenase b) Tyrosine
    c) Polyphenol oxidase d) None of these
47. Main aim of blanching is:
    a) To kill microorganism b) To inactivate enzyme
    c) To remove field heat d) None of these
48. Which of the following elements is responsible for transportation of anthocyanin pigments?
    a) P b) K
    c) Ca d) Al
49. Mineral element used to improve fruit quality is:
    a) Ca b) Mg
    c) Fe d) K
50. Which of the following is a non-PCR based molecular market?
    a) RAPD b) AFLP
    c) RFLP d) SSR

51. Oil content is the maturity indices of:
    a) Avocado b) Citrus
    c) Loquat d) Bael
52. Fruit cracking is not a severe problem in:
    a) Pomegranate b) Litchi
    c) Citrus d) All of these
53. Strawberry varieties grown respective of photoperiod are termed as:
    a) Short day b) Long day
    c) Short long day d) Ever bearing
54. Formation of black line at graft onion is a major problem in:
    a) Apple b) Almond
    c) Walnut d) Pecan nut
55. Sporophytic system of self incompatibility is observed in:
    a) Mango b) Aonla
    c) Both A and B d) Loquat
56. Presence of nucellar embryony is a major hurdle in breeding of:
    a) Apple b) Bael
    c) Citrus d) Papaya
57. A mutation which changes a mutant allele back into a wild type allele is known as:
    a) Forward mutation b) Reverse mutation
    c) Suppressor mutation d) Point mutation
58. Which of the following is a national mandarin variety?
    a) Coorg b) Kinnow
    c) Nagpur d) Khasi
59. Cultivated strawberry (*Fragaria* × *ananassa*) is a natural hybrid of:
    a) *Fragaria chiloensis* × *Fragaria vesca*
    b) *Fragaria chiloensis* × *Fragaria virginiana*
    c) *Fragaria virginiana* × *Fragaria vesca*
    d) *Fragaria virginiana* × *Fragaria chiloensis*
60. Variety of bananas is highly resistant to panama wilt disease:
    a) Basari b) Rajapuri
    c) Harichal d) Rajeli

61. The temperate fruit requiring maximum chilling is:
    a) Apple b) Pear
    c) Apricot d) Cherry

62. Which of the following temperate fruits requires minimum chilling?
    a) Almond b) Peach
    c) Plum d) Walnut

63. Which of the following is an indigenous variety of apple?
    a) Ambri b) Maharaji
    c) White Dotted Red d) Baldwin

64. Which of the following is a semi-dwarfing rootstock of apple?
    a) M.9 b) M.27
    c) MM.106 d) MM.104

65. A serious problem in mango hampering the export quality Alphonso is:
    a) Spongy tissue b) Jelly seed
    c) Internal fruit necrosis d) Black tip

66. Which of the following is major problem in greenhouse cultivation?
    a) White fly b) Aphid
    c) Nematode d) All of these

67. T-budding is a commercial method of propagation in:
    a) Mango b) Papaya
    c) Aonla d) Pomegranate

68. Gummosis and bark splitting are serious problems in:
    a) Jamun b) Aonla
    c) Citrus d) Jackfruit

69. The fruit type of litchi is:
    a) Berry b) Modified berry
    c) Aggregate d) Nut

70. The study of wine making is known as:
    a) Enology b) Virology
    c) Penology d) Vitology

71. In India, double pruning and double cropping systems in grape are followed in:
    a) Temperate region b) Hot tropical areas
    c) Mild tropical areas d) Sub-tropical region

72. Strawberry is commercially propagated by:

a) Suckers b) Runners

c) Tissue culture d) Stolen

73. Which of the following training systems of grape has high cost: benefit ratio?

a) Pergola system b) Kniffin system

c) Telephone system d) Head system

74. Which of the following fruit crops is not suitable for growing in subtropical climate?

a) European plum b) Asiatic plum

c) Peach d) Pear

75. The most stable sex form in papaya is:

a) Male b) Female

c) Hermaphrodite d) All of these

76. The presence of protandry and protogyny in the same crop is known as:

a) Dichogamy b) Heterodichogamy

c) Homogamy d) PDSD

77. Which of the following is a summer deciduous fruit crop?

a) Peach b) Loquat

c) Bael d) Ber

78. Which of the following is suitable for cultivation in degraded land?

a) Aonla b) Bael

c) Ber d) Custard apple

79. HCN in grape is used to:

a) Hasten early bud break b) Berry elongation

c) Increase fruitfulness d) Control vegetative growth

80. Total post-harvest losses in fruits amount to:

a) 15-20% b) 25-30%

c) 35-40% d) 45-50%

81. Regular severe pruning is practiced in:

a) Pear b) Plum

c) Peach d) Cherry

82. .... is a major physiological disorder of litchi.
    a) Sun burning b) Fruit cracking
    c) Pericarp browning d) All of these
83. The 'Spindle bush' system of training was given by :
    a) Dunn and Stolp b) Chalmer and Vandan Eude
    c) Beurill d) Schnitz Hubsh and Heinrichs
84. Temperate berries have originated from:
    a) North America b) South America
    c) Europe d) Asia
85. Which of following training systems is similar to central leader system but utilizes semi dwarfing rootstocks?
    a) Palmette leader b) Dwarf pyramid
    c) Spindle bush d) Solen system
86. The concept of 'Mega gene centres' was give by:
    a) N.I. Vavilov b) Harlan and de Wet
    c) Zeven and Zhukovsky d) Alphonse de Candolle
87. Temperature is a major limiting factor in cultivation of:
    a) Cashew nut b) Mango
    c) Litchi d) Loquat
88. Recently, seedless variety has been developed with fleshy stone in which of the following crop?
    a) Peach b) Plum
    c) Apricot d) Cherry
89. Canopy shape in central leader system is:
    a) Spherical b) Conical
    c) Columnar d) Pyramidal
90. Which type of training is known as 'bird's eye view'?
    a) Lincoln canopy b) Spindle bush
    c) Tatura trellis d) Modified central leader
91. Which of the following is an unauthorized IPR (Intellectual Property Right)?
    a) Copy right b) Trade mark
    c) Trade secret d) Geographical indication

92. Tetraploid citrus rootstocks have been developed for:
    a) Drought tolerance b) Salt tolerance
    c) Disease resistant d) Virus elimination
93. Which fruit crop is mainly grown in poly tunnel (protected condition) in India?
    a) Papaya b) Banana
    c) Strawberry d) Pineapple
94. Shifting of temperate fruit cultivation towards higher altitude occurs due to:
    a) Not meeting sufficient chilling requirement
    b) Increase in atmospheric CO_2 concentration
    c) Increased incidence of pest and diseases
    d) Higher summer temperature leads to cracking
95. The native place of Jamun is:
    a) China b) Malaysia
    c) India d) Bangladesh
96. Meristem culture is used to eliminate virus in which of the fruit crop(s)?
    a) Citrus b) Apple
    c) Pear d) All of these
97. Spacing followed in HDP of Amrapali variety of mango is:
    a) 2.5 m × 2.5 m b) 3 m ×2.5 m
    c) 2.5 m × 5.5 m d) 5.5 m ×5.5 m
98. Cashew nut is commercially propagated by:
    a) Wedge grafting b) Epicotyl grafting
    c) Tongue grafting d) Softwood grafting
99. Which of the following is a salt tolerant rootstock of citrus?
    a) *Citrus jambhiri* b) *Citrus Karna*
    c) *Citrus limonia* d) *Citrus aurantium*
100. Pruning in meadow orcharding of guava is done at:
    a) Once a year b) Twice a year
    c) Thrice a year d) Round the year
101. Bitterness in peach is due to:
    a) Amygdalin b) Prunasin
    c) Cyanogenic glycoside d) All of these

102. In India, meadow orcharding has been commercially adopted in:

a) Peach b) Pineapple

c) Grape d) Guava

103. Which of the following is used to overcome bud dormancy in grape?

a) HCN b) Ethrel

c) GA_3 d) Kinetin

104. High C:N ratio favours:

a) Vegetative growth b) Reproductive growth

c) Bud dormancy d) All of these

105. Recalcitrant is related with:

a) Seed germination b) Seed dormancy

c) Seed storage d) Seed production

106 percent pollinizers are recommended in apple orchard in India.

a) 11% b) 22%

c) 33% d) 44%

107. Which of the following is an immobile element in plant?

a) P b) Zn

c) Ca d) Fe

108. Excess application of nitrogen results in:

a) Improved fruit quality b) Deteriorate fruit quality

c) Fruit cracking d) Not effect fruit quality

109. Removal of low hanging branches leaving 7-10 cm in tree is referred to as:

a) Dehorning b) Ringing

c) Skirting d) Pollarding

110. The term 'Cincturing' refers to:

a) Removal of overcrowding and intermingling of branches

b) Practice of smoking the trees

c) Removal of a ring of bark out of the main trunk or branch of trees

d) All of these

111. Rest rootstock of citrus in south India is:

a) Cleopatra mandarin b) Citrange

c) Rangpur lime d) None of these

112. Running in mango is practiced during:
    a) May-June
    b) August-September
    c) November-December
    d) February-March
113. Seedlessness is a desirable character in improvement of:
    a) Mango
    b) Papaya
    c) Grape
    d) Banana
114. Little leaf is a physiological disorder of:
    a) Citrus
    b) Apple
    c) Grape
    d) All of these
115. Meristem tip culture is done to avoid:
    a) Insect infection
    b) Virus elimination
    c) Fungal infection
    d) Bacterial infection
116. Flat canopy is akin to which of the following training systems?
    a) Kniffin
    b) Cordon
    c) Solen
    d) Tatura trellis
117. Localized graft incompatibility between pear and quince can be overcome by:
    a) Insertion of mutually compatible interstock
    b) Removal of labile influences
    c) Removal of viral infection
    d) Insertion of compatible rootstock
118. Flowering occurs in autumn season in:
    a) *Prunus persica*
    b) *Prunus avium*
    c) *Prunus armeniaca*
    d) *Prunus communis*
119. Increase in soil pH in guava leads to:
    a) Guava wilt
    b) Viral disease
    c) Leaf bronzing
    d) Dieback
120. Recently commercialized bio-agent/ bio-fertilizer is:
    a) Phosphorus solubilizing
    b) Potassium solubilizing
    c) Vesicular arbuscular mycorrhiza
    d) Rhizobium

121. *Ex-situ* conservation restricts:

a) Gene flow b) Inbreeding
c) Disease and pest d) All of these

122. Mostly used protected structure by nurserymen in India is:

a) Plastic tunnel b) Poly house
c) Net house d) Glass house

123. The oldest training system followed in plum is:

a) Central leader b) Open centre
c) Modified central leader d) Tatura trellis

124. At present, nearly........................%age of the protected structure are being constructed by utilizing polythene sheet as a glazing material.

a) 20-40% b) 40-60%
c) 60-80% d) 80-90%

125. Polythene sheet is used as a covering material to prevent:

a) Ultra violet rays b) Humidity
c) Temperature d) CO_2

126. Which of the following factors is mostly maintained under green house:

a) Temperature and humidity b) Temperature and CO_2
c) Humidity and ventilation d) Light and CO_2

127. Silver coloured shade nets are used:

a) To destroy weeds b) To control aphids
c) To control fungal disease d) All of these

128. Lack of moisture causes:

a) Decrease in acidity b) Decrease in peel thickness
c) Decrease in firmness d) All of these

129. Premature leaf fall in apple is similar to the deficiency of which of the following nutrient?

a) Mg b) Ca
c) K d) N

130. First homozygous line was developed through androgenesis in which of these crop?

a) Citrus b) Apple
c) Pomegranate d) Grape

131. Bending operation in guava is done to facilitate:

a) Cultural operation b) Better fruiting

c) Early flowering d) All of these

132. Crop regulation (bahar treatment) in citrus and pomegranate is achieved by:

a) With holding of irrigation water

b) Root and shoot pruning

c) Deblossoming by application of NAA

d) All of these

133. How many buds are retained in rape after cane pruning?

a) 2-4 b) 4-6

c) 6-8 d) 8-12

134. Self-thinning property of New Perlette grape is a result of:

a) Chromosomal translocation b) Chromosomal transversion

c) Chromosomal doubling d) All of these

135. Serry shape in grape is controlled by:

a) Mono genetically b) Poly genetically

c) Cytogenetically d) None of these

136. Resistance to bacterial blight is governed byan allele of Daru.

a) Dominant b) Recessive

c) Additive d) None of these

137. Which of the following is/are triploid cultivar(s) of apple:

a) Baldwin b) Winesap

c) Gravenstin d) All of these

138. Which of the following PGR is used as substitute for chilling requirement in temperate fruits:

a) Kinetin b) Ethrel

c) GA_3 d) HCN

139. Which of the following types of incompatibility is observed in citrus?

a) Self-incompatibility b) Cross-incompatibility

c) Both self and cross d) None of these

140. For sugar translocation of grape, which of the following elements is necessary:

a) B　　b) Ca

c) Mg　　d) P

141. The pollination of fig is affected by:

a) House fly　　b) Wind

c) Honey bee　　d) Wasp

142. Acta Horticulturae is a publication of:

a) ASHS　　b) ISHS

c) JSHS　　d) HIS

143. Outer layer of papaya seed which hinders germination is called:

a) Integument　　b) Testa

c) Sarcotesta　　d) Embryo

144. Vegetatively propagated plants form a single mother plant which retains their characters after repeated multiplication are known as:

a) Clone　　b) Strain

c) Variety　　d) Cultivar

145. 'Mahal' a rootstock of pear is scientifically known as:

a) *Pyrus pyrifolia*　　b) *Pyrus pashia*

c) *Pyrus serotina*　　d) *Pyrus communis*

146. Which chemical is used for breaking bud dormancy in grapes?

a) Hydrogen cyanamide　　b) KNO3

c) Hydrogen peroxide　　d) All of these

147. Seed act was passed in India in the year:

a) 1966　　b) 1963

c) 1969　　d) 1968

148. Defoliation in Annona results in good crop. It is achieved by the spray of:

a) KSO4　　b) KCl

c) KIO3　　d) KI

149. Hexagonal system of planting accommodates .... (%) more plants than square system:

a) 10　　b) 15

c) 20　　d) 25

150. Predominant organic acid present in carambola is:

a) Citric b) Tartaric

c) Oxalic d) None of these

Answers Key

1.	(a)	2.	(c)	3.	(a)	4.	(b)	5.	(a)	6.	(c)	7.	(a)
8.	(a)	9.	(a)	10.	(a)	11.	(b)	12.	(a)	13.	(d)	14.	(a)
15.	(a)	16.	(b)	17.	(c)	18.	(a)	19.	(a)	20.	(a)	21.	(a)
22.	(c)	23.	(a)	24.	(b)	25.	(c)	26.	(b)	27.	(c)	28.	(b)
29.	(d)	30.	(a)	31.	(d)	32.	(d)	33.	(b)	34.	(a)	35.	(c)
36.	(b)	37.	(b)	38.	(a)	39.	(c)	40.	(a)	41.	(a)	42.	(a)
43.	(d)	44.	(b)	45.	(c)	46.	(c)	47.	(b)	48.	(d)	49.	(d)
50.	(c)	51.	(a)	52.	(c)	53.	(d)	54.	(c)	55.	(c)	56.	(c)
57.	(b)	58.	(c)	59.	(b)	60.	(a)	61.	(d)	62.	(a)	63.	(a)
64.	(c)	65.	(a)	66.	(c)	67.	(c)	68.	(c)	69.	(d)	70.	(a)
71.	(c)	72.	(b)	73.	(a)	74.	(a)	75.	(b)	76.	(b)	77.	(d)
78.	(a)	79.	(a)	80.	(b)	81.	(c)	82.	(b)	83.	(d)	84.	(a)
85.	(b)	86.	(c)	87.	(a)	88.	(a)	89.	(d)	90.	(c)	91.	(c)
92.	(b)	93.	(c)	94.	(a)	95.	(c)	96.	(d)	97.	(a)	98.	(d)
99.	(c)	100.	(c)	101.	(c)	102.	(d)	103.	(a)	104.	(b)	105.	(c)
106.	(c)	107.	(c)	108.	(b)	109.	(c)	110.	(c)	111.	(c)	112.	(b)
113.	(c)	114.	(d)	115.	(b)	116.	(d)	117.	(a)	118.	(b)	119.	(a)
120.	(b)	121.	(d)	122.	(a)	123.	(b)	124.	(d)	125.	(a)	126.	(b)
127.	(b)	128.	(d)	129.	(d)	130.	(a)	131.	(b)	132.	(d)	133.	(d)
134.	(b)	135.	(b)	136.	(b)	137.	(d)	138.	(c)	139.	(c)	140.	(a)
141.	(d)	142.	(b)	143.	(c)	144.	(a)	145.	(b)	146.	(a)	147.	(a)
148.	(d)	149.	(b)	150.	(c)								

12

ICAR – NET
Fruit Science Exam – 2017

1. 'National horticulture mission' was launched during
   a) 2003 b) 2004
   c) 2005 d) 2006
2. Agricultural and Processed Food Products Export Development Authority (APEDA) was established by the Government of India in:
   a) 1980 b) 1985
   c) 1990 d) 1995
3. Which of the following 'journal' is published from indian academy of horticultural sciences ?
   a) Indian journal of horticulture
   b) Progressive horticulture
   c) South indian journal of horticulture
   d) Indian journal of ornamental horticulture
4. '*Acta horticulturae*' is a publication of:
   a) ASHS b) ISHS
   c) JSHS d) HSI
5. Most suitable plant for bio-fencing is :
   a) Ker b) Karonda
   c) Phalsa d) Khijri
6. Removal of terminal portion of the shoots, branches or limbs leaving its basal portion is known as:
   a) Thinning out b) Heading back
   c) Skirting d) Dehorning
7. Multiple hedge row system is widely practiced in:
   a) Guava b) Peach
   c) Apple d) Cherry

8. Which of the following fruit crop can be grown at an altitude of 1500-2500 m msl ?
   a) Apple b) Aonla
   c) Ber d) Mango
9. Multiple row planting system is followed in:
   a) Apple b) Pear
   c) Peach d) Plum
10. Guava is traditionally planted at a spacing of :
    a) 3 x 3 m b) 6 x 6 m
    c) 9 x 9 m d) 12 x 12 m
11. Mango variety 'Pusa Pratibha' is generally planted at a spacing of:
    a) 2.5 x 2.5 m b) 6 x 6 m
    c) 8 x 10 m d) 10 x 10 m
12. Heavy pruning is practiced in:
    a) Citrus b) Aonla
    c) Phalsa d) Guava
13. The branching pattern where the secondary branches are in opposite side ?
    a) Monopodial - monochasium
    b) Monochasium - dichasium
    c) Sympodial - monochasium
    d) Sympodial - dichasium
14. Crown pinching in pineapple is done at:
    a) 25 daf b) 35 daf
    c) 45 daf d) 55 daf
15. Pruning in her under north indian condition is practiced in:
    a) March - April b) April - May
    c) May - June d) June - July
16. Which is the most suitable c/n ratio for growth and fruiting?
    a) 2 : 3 b) 1 : 3
    c) 3 : 2 d) 4 : 1
17. Apicial dominance is associated with .... hormone.
    a) Auxin b) Gibberellins
    c) Cytokinin d) Aba

18. Harvest index is the ratio of:
    a) Economic yield : biological yield
    b) Biological yield : economic yield
    c) Yield : evapo-transpiration
    d) None of these

19. Alternate bearing index value ranged from:
    a) 0 to 1 b) 1 to 10
    c) -1 to +1 d) 0 to s

20. Aternate bearing index value close to 0 means:
    a) Regular bearer b) Alternate bearer
    c) Moderate bearer d) Erratic

21. Emergence of new growth flush simultaneously with fruiting or immediately after harvest in mango is indication of:
    a) Regularity in bearing b) Precocity in bearing
    c) Dwarfness d) Drought tolerance

22. Smaller xylem vessel diameter and high bark percentage is the pre-selection criteria of:
    a) Regularity in bearing b) Precocity in bearing
    c) Dwarfness d) Vigorousness

23. Which of the following chemical used in walnut hulling?
    a) Ethrel b) Ga_3
    c) Iba d) Kinetin

24. Autonomic parthenocarpic ovary is rich in:
    a) Auxin b) Cytokinin
    c) Ethylene d) Gibberellin

25. In which of the fruit crops nucellar seedlings can be used to raise true-to-type plants?
    a) Sapota b) Guava
    c) Khirni d) Kagzi lime

26. The intervening period between the sequential emergences of leaves on the main stem of a plant is known as:
    a) Phyllochron b) Dichotomy
    c) Both d) None of the above

27. Leaf pruning is practiced in:
   a) Date palm b) Custard apple
   c) Sapota d) Titchi
28. The phenomenon of 'double dormancy[9] exists due to:
   a) Rest embryo + impermeable coat
   b) Rest embryo + permeable coat
   c) Environment + impermeable coat
   d) Environment + permeable coat
29. Vitamin d is chemically known as:
   a) Alpha-tocopherol b) Calciferol
   c) Riboflavin d) Retinol
30. The phenomenon in which different parts of a plant show phase variation and if propagated through meristems, the different phase are perpetuated in the offspring is known as:
   a) Topophysis b) Periphysis
   c) Neophysis d) None of these
31. Which of the following mechanism(s) is (are) involved in salt tolerance in plants?
   a) High k/na ratio b) Salt extrusion
   c) Accumulation of solute d) All the above
32. Which of the following enzyme is responsible for colouration of living tissues in tz test?
   a) Dehydrogenase b) Epoxigenase
   c) Esterase d) Dismutase
33. Mottle leaf of citrus is due to the deficiency of:
   a) Mg b) Fe
   c) Zn d) B
34. When a root grows in the direction of the force of gravity (downwards) is known as:
   a) Positive gravitropism b) Negative gravitropism
   c) Positive phototropism d) Negative phototropism
35. A seed that will not germinate unless it is exposed to light is known as:
   a) Photodormant b) Thermodormant
   c) Ecodormant d) Thermo insensitive

36. Chilling requirement of most the temperate fruits is:
    a) Below 7 b) Above 7
    c) Below 0 d) Above 0
37. Which of the following is the most common chilling model ?
    a) Winter chilling model b) Utah chilling model
    c) Dynamic chilling model d) Shimla chilling model
38. Low temperature treatment of seeds to promote germination is called as;
    a) Scarification b) Stratification
    c) Seed priming d) None of the above
39. Chemical used for controlling sprouting of onions is:
    a) Maleic hydrazide b) Ethylene
    c) Gibberellic acid d) Cytokinin
40 Which of the following chemical is used for polyploidy breeding?
    a) Colchicine b) Mh
    c) Naa d) Sadh
41. The *Avena* (oat) coleoptile curvature test is done for:
    a) Auxin b) Cytokinin
    c) Gibberellins d) Ethylene
42. Which of the following classification is not correctly matched?
    a) Citrus : tanaka and swingle
    b) Mango : kosterman and, bompard
    c) Banana : simmonds and shepherd
    d) Grape : kuhn
43. Ideal characteristics for mango rootstock breeding is/are:
    a) Dwarfing b) Polyembryonic
    c) Resistant to biotic and abiotic stresses d) All the above
44. Morphological characters used for mango classification is/are:
    a) Floral disc, fertile stamens b) Leaf venation
    c) Shape and colour of inflorescence d) Peel and pulp colour

45. Which of the following is an indigenous species of mango?
    a) *Mangifera camptosperma*
    b) *Mangifera odorata*
    c) *Mangifera pajang*
    d) *Mangifera macrocarpa*
46. Which of the following *mangifera* species has labyrinthine seed?
    a) *Mangifera griffithi*
    b) *Mangifera monandra*
    c) *Mangifera paludosa*
    d) *Mangifera gedebe*
47. Mango hybrid mallika is a product of cross between:
    a) Neclam x dushehari
    b) Dushehari x neelam
    c) Alphonso x neelam
    d) Neelam x alphonso
48. Which of the following mango variety has wider adaptability ?
    a) Malllika
    b) Alphonso
    c) Kesar
    d) Zardalu
49. The secondary centre of origin of mango is:
    a) Myanmar
    b) Western Malaysia
    c) Florida
    d) New Guinea
50. Based on maturity 'banganapalli' mango is classified as:
    a) Early
    b) Mid
    c) Late
    d) Very late
51. The aez (agri export zone) identified for kesar mango is:
    a) Andhra Pradesh and Maharashtra
    b) Gujarat and Maharashtra
    c) Maharashtra and Karnataka
    d) Gujarat and Uttar Pradesh
52. 'Lakhi bagh' is associated with the planting of:
    a) Litchi
    b) Mango
    c) Citru
    d) Guava
53. Which of the following is not correctly matched?
    a) Pusa surya = eldon x amrapaii
    b) Arka suprabhat = amrapaii x arka anmol
    c) Ambika = amrapali x anardnn pasand
    d) Konkan ruchi = neclum x alphonso

54. 'Arunika' mango was released from:

a) CISH, Lucknow | b) BSKKV, Dapoli
c) FRS, Sangareddy | d) MPKV, Rahuri

55. Creeping variety of mango is:

a) Dwarf | b) Semi-dwarf
c) Vigorous | d) Semi-vigorous

56. Mulching in mango is practiced to check:

a) Transpiration | b) Evaporation
c) Disease and pests | d) Nematodes

57. Which of the following is a monocotyledonous, monoecious, monocarpic, mesophytic, herbaceous perennial?

a) Pineapple | b) Papaya
c) Litchi | d) Banana

58. Tightly clump forming banana species is:

a) *Musa balbisiana* | b) *Musa acuminata*
c) *Musa textilis* | d) *Musa basjoo*

59. Novaria a variety of banana has been developed using:

a) Polypoidy | b) *In vitro* mutagenesis
c) Hybridization | d) Embryo rescue

60. Genome constitution of 'popoulu' banana is:

a) AAB | b) AAA
c) ABB | d) ABBB

61. Days required from full bloom to maturity for 'Dwarf Cavendish' banana is:

a) 90 ± 5 | b) 100 ± 5
c) 125 ± 5 | d) 130 ± 5

62. The storage temperature of banana is:

a) 10°C | b) 13°C
c) 15°C | d) 8°C

63. Smoking is used for ripening of:

a) Mango | b) Banana
c) Pineapple | d) Papaya

64. Main obstacle in banana breeding is:
    a) Male sterility and parthenocarpy
    b) Female sterility
    c) Triploidy
    d) Polyembryony
65. Papaya leaf curl virus is transmitted by :
    a) Aphid b) Jassid
    c) White fly d) Hopper
66. Which of the following is/are lethal combination in papaya?
    a) MM b) M^hM^h
    c) MM^h d) All of these
67. Mountain papaya is:
    a) *Carica parviflora* b) *Carica candamarcensis*
    c) *Carica papaya* d) *Carica microcarpa*
68. Pineapple variety suitable for canning is:
    a) Kew b) Jnidhup
    c) Lakhat d) Queen
69. Tctrasomic rootstock of guava is:
    a) Pusa srijan b) *P. molle*
    c) 110 R d) Lalit
70. Which grape variety is commercially used for champagne making in the world ?
    a) White riesling b) Thompson seedless
    c) Pinot noir d) Pearl of casaba
71. 'Arkavati' is a hybrid of:
    a) Black champa x thompson seedless
    b) Black champa x anab-e-shahi
    c) Black champa x queen of vineyard
    d) Black champa x beauty seedless
72. Shot berry formation is a major problem in which of the following grape variety:
    a) Perlette b) Thompson seedless
    c) Anab-e-shahi d) Bangalore blue

73. Anthesis in grapes occurs at:

a) 6-7 am b) 7-8 am

c) 8-9 am d) 9-10 am

74. Embryo rescue in grape is used for the production of :

a) Seeded x seeded b) Seeded x seedless

c) Seedless x seedless d) All the above

75. King tangor x owari satsuma are parentage of:

a) Kara b) Wilking

c) Orlando d) Murcott

76. Interspecific hybrids are most successful in:

a) Citrus b) Mango

c) Banana d) Grape

77. Sessile flowers on previous season wood with clamp like structure are found in:

a) *Fortunella hindsii* b) *Eremocitrus glauca*

c) *Fortunella japonica* d) *Poncirus trifoliate*

78. Which is a cl excluder rootstock in citrus?

a) Sour orange b) Trifoliate orange

c) Rangpur lime d) Rough lemon

79. Characteristic feature of 'flying dragon[9] citrus rootstock is:

a) Dwarfing b) Vigorous

c) Semi-vigorous d) Gigantic

80. Indicator plant for xyloporosis in citrus is:

a) Etrong citron b) Mexican lime

c) Sweet lime d) Rangpur lime

81. Which is used to induce dwarfing in citrus?

a) Virus b) Bacteria

c) Viroid's d) Mycoplasma

82. 'Tangor' is an inter-specific cross between:

a) C. *reticulata* x C. *aurantiifolia*

b) C. *reticulata* x C. *paradisi*

c) C. *reticulata* x C. *sinensis*

d) C. *reiiculata* x C. *aurantium*

83. Resistance to bacterial blight in 'Daru' pomegranate is governed by:
    a) Dominant genes
    b) Recessive genes
    c) Additive genes
    d) None of these
84. 'Gul-e-Sbah Rose Pink' in pomegranate breeding used as donor parent for
    a) Dwarfness
    b) Resistant to blight
    c) Mellowness
    d) Dark aril colour
85. Which of the following fruit has narrow genetic diversity?
    a) Mangosteen
    b) Litchi
    c) Longan
    d) Rambutan
86. Which of the following is a non-climacteric fruit?
    a) Apricot
    b) Bael
    c) Rambutan
    d) Pear
87. Avocado flowers are:
    a) Morphologically bisexual and functionally unisexual
    b) Functionally bisexual and morphologically unisexual
    c) Morphologically and functionally unisexual
    d) Morphologically and functionally bisexual
88. 'Arka Sahan' custard apple is a progeny of:
    a) *A. atemoya* x *A. squamosa*
    b) *A. atemoya* x *A. cherimola*
    c) *A. atemoya* x *A. reticulata*
    d) *A. squamosa* x *A. atemoya*
89. Drought tolerant species of Annona is:
    a) *Annona glabra*
    b) *Annona cherimola*
    c) *Annona reticulata*
    d) *Annona atemoya*
90. The term 'pcna' is related to:
    a) Persimmon
    b) Passion fruit
    c) Litchi
    d) Durian
91. Cucumber tree is known as:
    a) Bilimbi
    b) Jungli jalebi
    c) Lotka
    d) Buddha's hand
92. Gisela and \veiroot are popular rootstocks of:
    a) Pear
    b) Cherry
    c) Plum
    d) Peach

93. Commonly used cherry rootstock in india is:

a) Gisela b) Colt
c) Mazzard d) Mahaleb

94. Delicious apple in india was introduced by:

a) S. N. stokes b) W. T. Scott
c) Captain lee d) Luther burbank

95. For harvesting delicious apple at right maturity, the starch pattern index (spi) should be:

a) 2.5/10 b) 3.5/10
c) 4.5/10 d) 6.5/10

96. Akbar* an apple hybrid is a cross of:

a) Sunheri x prima b) Red delicious x ambri
c) Ambri x cox orange pippins d) Golden delicious x rome beauty

97. San Josh scale of apple was introduced in India from:

a) UK in 1906 b) USA in 1936
c) France in 1916 d) France in 1906

98. Most widely used dwarfing rootstock of apple in the world is:

a) M 9 b) M 27
c) B 9 d) P 22

99. Which of the following series of apple rootstock is resistant to viruses?

a) MI series b) MM series
c) EMLA series d) All of these

100. 'Waterloo' walnut was introduced in india from:

a) France b) USA
c) Spain d) Australia

101. 'Mahan' is a variety of:

a) Walnut b) Pecan nut
c) Apricot d) Almond

102. 'Albinism' is a disorder of:

a) Apple b) Almond
c) Apricot d) Strawberry

103. 'Mouse ear' is a physiological disorder of:

a) Pecan nut b) Pistachio nut
c) Chestnut d) Hazelnut

104. Edible part of loquat is:

a) Fleshy receptacle
b) Fleshy thalamus
c) Fleshy aril
d) Fleshy peduncle

105. Type of incompatibility reported in loquat is:

a) Sporophytic
b) Gametophytic
c) Genotypic
d) All of these

106. Nectarine is fuzzless mutant of:

a) Plum
b) Almond
c) Peach
d) Apricot

107. The Sab-family of *Eriobotrya japonics* is:

a) Pomoideae
b) Prunoideae
c) Rosoideae
c) Citroideae

108. Which of the following country has highest area under protected cultivation ?

a) Israel
b) Japan
c) Usa
d) China

109. The fruit crop which is mostly cultivated under protected cultivation in india is :

a) Banana
b) Papaya
c) Pineapple
d) Strawberry

110. A major challenge in the protected cultivation of fruit crops is the increasing threat of :

a) White flies
b) Aphids
c) Fruit fly
d) Nematodes

111. Cocopeat has .... per cent porosity.

a) 15-20%
b) 20-25%
c) 25-30%
d) 30-35%

112. Expanded form of cfb is:

a) Coloured fiber board
b) Corrugated filler box
c) Corrugated fiber board
d) Complete filling box

113. Which of the following attained highest TSS at ripening ?

a) Mango
b) Bael
c) Aonla
d) Papaya

114. Browning in ripe fruit due to PPO oxidation can be prevented by:

a) Ascorhate b) Sugar

c) Protein d) All of these

115. Dehydration of fruit crops is done at :

a) 40-50°C b) 50-60°C

c) 60-70°C d) 70-80°C

116. Most commonly used dryer for drying of fruit juices is:

a) Kiln dryer b) Cabinet dryer

c) Rotary dryer d) Spray dyer

117. Vapour heat treatment in mango is done at :

a) 47.5 °c for 5 minutes b) 47.5 °c for 20 minutes

c) 46.5 °c for 5 minutes d) 45.5 °c for 20 minutes

118. Father of modern refrigeration is:

a) John Gorrie b) Paul Berg

c) James Harrison d) Keith Oliver

119. The quality of papain (papaya latex) is measured by:

a) Methionine b) Tyrosine

c) Lysine d) None of these

120. Desapping is practiced in :

a) Mango b) Litchi

c) Sapota d) Jackfruit

121. Pectin is measured by :

a) Thermometer b) Refractometer

c) Jelmeter d) All of these

122. Important constituent of jelly is:

a) Sugar b) Acid

c) Water d) Pectin

123. Headquarter of ciphet is located at :

a) Mysore b) Ludhiana

c) Kundli d) Bhopal

124. In which fruit crop, tissue culture is commercially exploited ?

a) Apple b) Mango

c) Aonla d) Banana

125. Which of the following is a co-dominant marker ?
   a) RAPD b) SSR
   c) ISSR d) AFLP
126. Gwas stands for :
   a) Genome-wide association study
   b) Guanine wide association study
   c) Genuine water association study
   d) None of these
127. Markers used for diversity analysis:
   a) Monomorphic b) Polymorphic
   c) Heteromorphic d) All of these
128. National Certification System for Tissue Culture raised plants (NCS-TCP) is being implemented by:
   a) ICAR b) DBT
   c) CSIR d) DST
129. Explant used for somatic embryogenesis in mango:
   a) Pollen b) Ovule
   c) Nucellus d) Embryo
130. Plants containing nucleus of one species but cytoplasm from both the parental species are known as:
   a) Somatic hybrids b) Cybrids
   c) Hybrids d) Haploids
131. Widdy used gene transfer technology in fruit crops is:
   a) Agrobacterium-mediated b) PEG-mediated
   c) Silicon carbide fiber-mediated d) Particle bombardment
132. Croup of organism crossed freely and produce fertile progeny is:
   a) Species b) Population
   c) Ecosystem d) Biome
133. The populations of all the different species that live together in an ecosystem is :
   a) Community b) Population
   c) Biome d) Ecosystem

134. Primary data collected at the time of germplasm entry is:

a) Passport data
b) Accession number
c) Indigenous collection
d) Exotic collection

135. Cryopreservation is associated with

a) Liquid nitrogen
b) Liquid oxygen
c) Liquid potassium
d) Liquid carbon dioxide

136. Triploids varieties of guava are generally :

a) Seeded
b) Seedless
c) Partially seeded
d) Partially seedless

137. Grape species resistant to downy mildew and powdery mildew are:

a) V. *amurensis and V. aestivalis*
b) V. *omurensis and V. parviflora*
c) V. aestivalis and *V. rotundifolia*
d) V. aestivalis and V. candicans

138. Mandarin and sweet orange hybrid is:

a) Calamondin
b) Cleopatra
c) Clementine
d) All of these

139. The purple passion fruit varieties and their hybrids arc :

a) Self-compatible
b) Self-incompatible
c) Partially compatible
d) Partially incompatible

140. Which of the following statement is true about prunes ?

a) All plums are prunes
b) All prunes are plums
c) All peaches are prunes
d) All prunes are peaches

141. What is the use of irradiation in food preservation :

a) Extending the shelf life of food
b) Delay of ripening
c) Control of insects pests and disease
d) All of these

142. Which of the following is/are advantage of protected cultivation?

a) Early harvesting
b) Off season production
c) Higher yield and improved quality
d) All of these

143. Irrigation system commonly used in protected cultivation:

a) Drip irrigation
b) Sprinkler irrigation
c) Sub-surface irrigation
d) Check basin irrigation

144. India has joined hands uith which country for hi tech nursery technology for higher productivity?

a) China
b) Japan
c) Israel
d) Australia

145. Main objective of ventilation storage structure is/are

a) Management of temp. & humidity
b) To supply fresh air for respiration
c) To reduce pests and diseases
d) All of these

Answers Key

1.	(c)	2.	(b)	3.	(a)	4.	(b)	5.	(b)	6.	(b)	7.	(c)
8.	(a)	9.	(a)	10.	(b)	11.	(b)	12.	(c)	13.	(d)	14.	(c)
15.	(c)	16.	(c)	17.	(a)	18.	(a)	19.	(a)	20.	(a)	21.	(a)
22.	(c)	23.	(a)	24.	(a)	25.	(d)	26.	(a)	27.	(a)	28.	(a)
29.	(b)	30.	(a)	31.	(d)	32.	(a)	33.	(c)	34.	(a)	35.	(a)
36.	(a)	37.	(a)	38.	(b)	39.	(b)	40.	(a)	41.	(a)	42.	(d)
43.	(d)	44.	(a)	45.	(a)	46.	(d)	47.	(a)	48.	(a)	49.	(c)
50.	(a)	51.	(b)	52.	(b)	53.	(a)	54.	(a)	55.	(a)	56.	(b)
57.	(d)	58.	(a)	59.	(b)	60.	(a)	61.	(a)	62.	(b)	63.	(b)
64.	(a)	65.	(c)	66.	(d)	67.	(b)	68.	(a)	69.	(a)	70.	(a)
71.	(a)	72.	(a)	73.	(b)	74.	(c)	75.	(a)	76.	(a)	77.	(d)
78.	(c)	79.	(a)	80.	(c)	81.	(c)	82.	(c)	83.	(b)	84.	(d)
85.	(a)	86.	(c)	87.	(a)	88.	(a)	89.	(c)	90.	(a)	91.	(a)
92.	(b)	93.	(c)	94.	(a)	95.	(c)	96.	(c)	97.	(d)	98.	(a)
99.	(c)	100.	(b)	101.	(b)	102.	(d)	103.	(a)	104.	(b)	105.	(b)
106.	(c)	107.	(a)	108.	(d)	109.	(d)	110.	(d)	111.	(c)	112.	(c)
113.	(b)	114.	(a)	115.	(c)	116.	(b)	117.	(b)	118.	(c)	119.	(b)
120.	(a)	121.	(c)	122.	(d)	123.	(b)	124.	(d)	125.	(b)	126.	(a)
127.	(b)	128.	(b)	129.	(c)	130.	(b)	131.	(a)	132.	(a)	133.	(a)
134.	(a)	135.	(a)	136.	(b)	137.	(a)	138.	(c)	139.	(a)	140.	(b)
141.	(d)	142.	(d)	143.	(a)	144.	(c)	145.	(d)				

13

ICAR – NET Fruit Science Exam – 2018

1. The average productivity of fruits in india is:
   a) 9.2 t/ha b) 12.2 t/ha
   c) 14.2 t/ha d) 16.2 t/ha
2. The leading state in fruit production in india is:
   a) A.P. b) U.P.
   c) Maharashtra d) Karnataka
3. The leading state in pineapple production in india is:
   a) Maharashtra b) Assam
   c) West Bengal d) Tripura
4. The top producer country of apple in the world is:
   a) China b) Italy
   c) United states d) Spain
5. 'Albinism' is a physiological disorder of:
   a) Banana b) Pineapple
   c) Strawberry d) Papaya
6. Pusa Surya' is an improved variety of:
   a) Papaya b) Grape
   b) Guava d) Mango
7. Plum is botanically known
   a) *prunus cerasifera* b) *prunus domestica*
   c) *prunus salicina* d) *prunus americana*
8. 'Hard end' is a physiological disorder of:
   a) Peach b) Pear
   c) Plum d) Apple
9. Indicator plant for citrus greening is:
   a) Mosambi b) Valencia
   c) Jaffa d) Hamlin

10. 'Safed jam' variety of guava is a cross between:
    a) Allahabad safeda x Kohir
    b) Kohir x Allahabad Safeda
    c) Allahabad Safeda x Apple colour
    d) Seedless x Allahabad Safeda
11. Amrapali x Senasation are the parents of:
    a) Pusa Arunima b) Pusa Pratibha
    c) Pusa Shrestha d) All of these
12. Which of the following fruit crop (s) is/are considered as manmade fruits?
    a) Kinnow b) Strawberry
    c) Atemoya d) All of these
13. Which of the following fruit crop is referred as strictly subtropical owing to its exact climatic requirement?
    a) Litchi b) Olive
    c) Loquat d) Persimmon
14. Wood apple belongs to the family:
    a) Rosaceae b) Ryrtaceae
    c) Rutaceae d) Rapindaceae
15. The percentage of iron in dry karonda is about:
    a) 40% b) 20%
    c) 28.5% d) 39.1%
16. Type of self-incompatibility reported in mango is:
    a) Sporophytic b) Gametophytic
    c) Both d) None of these
17. Seedlessness in grape is controlled by:
    a) Single dominant gene b) Single recessive gene
    c) Double dominant gene d) Double recessive gene
18. Seedlessness in lemon *(citrus limori)* is due to:
    a) Self-incompatibility b) Parthenocarpy
    c) Ovule sterility d) Embryo abortion
19. Shoot tip grafting has been commercialized in india in:
    a) Sweet orange b) Mandarin
    c) Grapefruit d) Rangpur lime

20. 'Winter banana[9] is a variety of:
    a) Mango b) Apple
    c) Banana d) Pear
21. *Annona squamosa* is native to:
    a) Peru b) West indies
    c) Bolivia d) Paraguay
22. 'Punjab beauty' is a variety of:
    a) Peach b) Plum
    c) Pear d) Walnut
23. Triple sigmoidal growth curve is observed in:
    a) Fig b) Kiwifruit
    c) Aonla d) Ber
24. Inverted bottleneck is occur in her when budded onto:
    a) *Ziziphus rotundifolia* b) *Ziziphus paludosa*
    c) *Ziziphus onepilea* d) *Ziziphus nummularia*
25. The best intercrop crop planted in banana orchard is:
    a) Papaya b) Pineapple
    c) Ginger d) Pepper
26. 'Notching' is practiced in:
    a) Fig b) Wood apple
    c) Jamun d) Grape
27. The office of 'Geographical indications' is located at:
    a) New Delhi b) Mumbai
    c) Chennai d) Kolkata
28. Protection of Plant Varieties and Farmers' Rights (PPV&FR) Authority is located at:
    a) New delhi b) Bangalore
    c) Lucknow d) Mumbai
29. Aroma of most of the fruits are due to:
    a) Terpenes b) Esters
    c) Lactone d) Acetone
30. Which of the following is a suitable crop for western arid himalayan region?
    a) Ber b) Aonla
    c) Walnut d) Bael

31. Day length does not affect flowering in
    a) Pineapple b) Passion fruit
    c) Mango d) Papaya
32. Which of the following is an orinthophillus (birds pollinated) fruit crop?
    a) Pineapple b) Banana
    c) Strawberry d) Oil palm
33. The edible part of fig is:
    a) Pericarp
    b) Thalamus
    c) Fleshy receptacles and placenta
    d) Bracts and perianth
34. Cauliflory is observed in:
    a) Jackfruit b) Peach
    c) Loquat d) All of these
35. Protogyny is found in which fruit crop?
    a) Sapota b) Walnut
    c) Passion fruit d) Coconut
36. 'Ganesh kirti' is a variety of:
    a) Ber b) Bael
    c) Pomegranate d) Aonla
37. Which of the following is a variety of bael?
    a) Goma kirti b) Goma yashi
    c) Goma priya d) Goma aishwarya
38. 'Thar Sevika' is a cross of :
    a) Seb x Katha b) Seb x Jogia
    c) Seb x Umran d) Illaichi x Katha
39. 'Sonaka' is a clonal selection of:
    a) Kishmish chaurni b) Anab-e-shahi
    c) Cheema sahebi d) Thompson seedless
40. 'Arka Hans' is a variety of:
    a) Mango b) Guava
    c) Custard apple d) Grape

41. Which of the following is a spur pruning grape variety?
    a) Perlette b) Anab-e-shahi
    c) Thompson seedless d) Bhokri
42. Optimum temperature required for flowering in pineapple is:
    a) 15-20°C b) 20-25°C
    c) 25-30°C d) 30-35°C
43. Which of the following training system of grape is not practiced in india?
    a) Bower b) Telephone
    c) Kniffin d) Pergola
44. Seedless mango cv. sindhu is a result of back cross between:
    a) Neelum x alphonso b) Ratna x alphonso
    c) Banganpalli x alphonso d) Alphonso x totapuri
45. Which of the following rootstock imparts dwarfness in alphonso mango?
    a) Moovandan b) Kurukkan
    c) Nekkare d) Vellaicolomban
46. Double sipmoid growth curve is observed in:
    a) Pear b) Apple
    c) Plum d) Almond
47. Which of the following is a dwarf variety of sapota?
    a) PKM-1 b) PKM-2
    c) PKM-3 d) DSH-1
48. India has greater diversity of:
    a) Cashew b) Citrus
    c) Guava d) Avocado
49. Aggregate fruits are developed from:
    a) Single flower b) 2-3 flowers
    c) Many flowers d) All of these
50. Which of the following produce seedless fruits due to partheno-carpy?
    a) Capri fig b) Mango
    c) Seedless guava d) Grape
51. 'Mazzard' is a rootstock of:
    a) Sour cherry b) Sweet cherry
    c) Plum d) Pear

52. Sugar content in finished jelly should be:
    a) 70% b) 45%
    c) 35% d) 65%
53. Fruit which is used for preparation of good quality jelly is:
    a) Mango b) Jamun
    c) Aonla d) Guava
54. Exhaustion a process involved in:
    a) Canning b) Jelly making
    c) Pickling d) None of these
55. For canning of mango slices, the TSS of sugar solution should be-
    a) 20°B b) 30°B
    c) 40°B d) 50°B
56. Premature pollen germination causes a low fruit set in:
    a) Persimmon b) Plum
    c) Walnut d) Pecan nuts
57. Pin type of heterostyly is observed in:
    a) Peach b) Plum
    c) Pomegranate d) Carambola
58. Calyx cavity dehiscence is a disorder of:
    a) Persimmon b) Plum
    c) Papaya d) Peach
59. Which of the following fruit crop can be successfully grown under protected environment?
    a) Kiwifruit b) Grape
    c) Kokum d) Passion fruit
60. Fruit crop having highest area under protected cultivation is:
    a) Strawberry b) Grape
    c) Banana d) Papaya
61. Which of the following is not a variety of guava?
    a) Arka mridula b) Arka amulya
    c) Arka shyam d) Arka kiran
62. Which of the following is most susceptible to water logging?
    a) Banana b) Mango
    c) Papaya d) Guava

63. Which of the following is the most dwarfing rootstock of citrus?
    a) Rangpur lime   b) Troycr citrange
    c) Flying dragon   d) Sour orange
64. Which of the following rootstock provides the lowest height to apple?
    a) M 9   b) M 27
    c) MM 111   d) MM 104
65. To provide proper shape and form to the canopy, which of the following practice is followed?
    a) Training   b) Pruning
    c) Chemical approach   d) All of these
66. Crop regulation in guava can be achieved by:
    a) With holding of irrigation   b) Deblossoming
    c) Root pruning   d) All of these
67. 'Hen and chicken' in grapes occurs due to the deficiency of:
    a) Ca   b) Fe
    c) B   d) K
68. Which of the following is a highly mobile element in plants?
    a) K   b) Mn
    c) Ca   d) Zn
69. Which of the following is not an essential element to the plants?
    a) N   b) Si
    c) Ni   d) Mo
70. 'Apple colour' is a variety of:
    a) Apple   b) Guava
    c) Mango   d) Papaya
71. Frost injury is a major problem in:
    a) Central leader system   b) Modified leader system
    c) Open centre   d) All of these
72. The dwarfing rootstock of pear is:
    a) Quince A   b) Quince B
    c) Quince C   d) All of these
73. 'Fuerte' is a leading cultivar of:
    a) Avocado   b) Durian
    c) Loquat   d) Persimmon

74. Which of the following is viviparous in nature?
    a) Papaya b) Jackfruit
    c) Sapota d) Loquat
75. Papaya variety resistant to ring spot virus is:
    a) Honey gold b) Solo
    c) Red lady d) Sunup
76. In date palm, which of the following condition occurs:
    a) Mataxenia and monoecy b) Mataxenia and dioecy
    c) Xenia and monoecy d) Xenia and dioecy
77. In aonla, the male flowers first appear in cluster in the ....of leaves on lower part of the determinate shoot.
    a) Basal b) Tip
    c) Axil d) All of these
78. Prunacin is present in:
    a) Pear b) Peach
    c) Plum d) Walnut
79. High temperature and high humidity in grape orchard favours:
    a) Early maturity b) Early bud break
    c) Downy mildew d) Powdery mildew
80. Which of the following is a growth retardant?
    a) Auxin b) Gibberellin
    c) Ethylene d) Cultar
81. Abscisic acid (ABA) is synthesized in:
    a) New leaves b) Older leaves
    c) Buds d) Flowers
82. Which of the following plant part have highest totipotency?
    a) Xylem b) Phloem
    c) Cambium d) Cortex
83. Cleistogamy is observed in:
    a) Grape b) Citrus
    c) Sapota d) None of these

84. Major colouring compound of the banana fruit skin is:
   a) Carotenoids b) Anthocyanin
   c) Lycopene d) Chlorophyll

85. Which of the following is a drought tolerant banana?
   a) Culcutta-4 b) SH-3149
   c) Musa balbisiana d) All Of these

86. In Musa sapientum, the male bud is:
   a) Persistent b) Non persistent
   c) Both d) None of these

87. Closer spacing in orchard creates a microclimate around the canopy, which includes?
   a) Increasing of temperature, light and relative humidity
   b) Decreasing of temperature, light and relative humidity
   c) Decrease of temperature and light, increase of relative humidity
   d) Increase of temperature and decrease of light & relative humidity

88. To improve light penetration in banana plantation, which of the following is practiced?
   a) Removal of surplus leaves b) Tying of functional leaves
   c) Desuckering d) All of these

89. Phytochrome is related to:
   a) Seed germination b) Transpiration
   c) Flowering d) All of these

90. A population in which every plant has equal chance to pollinate others is known as:
   a) Random mating b) Open pollinated
   c) Close mating d) Cross pollinated

91. For uniform light distribution in the greenhouse which of the following glazing material should be used?
   a) Diffused plastic film b) Clear plastic film
   c) Diffused glass d) None of these

92. Which of the following is a temporary preservation method?
   a) Pasturization b) Sterilization
   c) Canning d) Fermentation

93. Blanching of fruits is done at:
    a) 80°C b) 90°C
    c) 100°C d) 120°C
94. Dehydration of fruit crops is done at:
    a) 40-50°C b) 50-60°C
    c) 60-70°C d) 70-80°C
95. Benzoic acid is effective against:
    a) Fungi b) Bacteria
    c) Yeast d) Mould
96. TSS of RTS is:
    a) 10°B b) 20°B
    c) 30°B d) 40°B
97. Which of the following is known as Ripening Hormone'?
    a) Auxin b) Cytokinin
    c) Gibberellins d) Ethylene
98. In which of the following fruit CAM cycle is found:
    a) Strawberry c) Pineapple
    b) Papaya d) Sapota
99. Apical dominance is associated with..............hormone.
    a) Auxin b) Gibberellins
    c) Cytokinin d) ABA
100. The major obstacles in adoption of precision farming in India is/are:
    a) Small land holding b) Poor economic status of farmers
    c) Lack of success stories d) All of these
101. Which of the following is a cytokinin?
    a) BA b) ABA
    c) IBA d) 2,4-D
102. Which of the following is/are subtropical peaches:
    a) Shan-e-Punjab b) Sharbati
    c) Florda Sun d) All of these
103. Which of the following affects flowering?
    a) Plant age b) Temperature
    c) Nutrient status d) Wind

104. "Tropism' is controlled by:

a) Auxin b) Cytokinin

c) Gibberellins d) Ethylene

105. Which of the following fruit is not suitable for jam making?

a) Mango c) Aonla

b) Banana d) Lemon

106. Which of the following fruit has narrow genetic diversity in India?

a) Sapota c) Phalsa

b) Jackfruit d) Banana

107. The fruit crop which has greater biodiversity in India is:

a) Olive b) Kiwifruit

d) Apricot c) Peach

108. Skin cracking is commonly observed in:

a) Grape b) Litchi

c) Bael d) All of these

109. Exhausting is related to:

a) Fermentation b) Canning

c) Sterilization d) Dehydration

110. Which of the following is an indigenous variety of pineapple?

a) Kew b) Queen

c) Spanish d) Lakhat

111. The diversity of fruit crops comprises of:

a) Landraces and local selections b) Wild and weedy species

c) Elite cultivars d) All of these

112. Which of the following is not a pome fruit?

a) Apple c) Plum

b) Pear d) Loquat

113. A light or thin fog is usually called as:

a) Dew b) Mist

c) Smoke d) Frost

114. Which of the following is/are main cause(s) of low productivity of mango in India?

a) Old and senile orchard b) Low density of planting

c) Lack of quality planting material d) All of these

115. Which of the following is not division method of propagation?

a) Rhizome a
c) Tuberous roots
b) Tuber
d) Bulb

116. Which of the following method is practiced during the period when there is lack of sap flow and bud does not slip out easily from the bark?

a) Chip budding
b) Ring budding
c) Forkert budding
d) Patch budding

117. A thin and long stem which develops from the axil of leaves is known as:

a) Suckers
c) Runners
b) Stolen
d) Offshoots

118. 'Allison' is a variety of:

a) Strawberry
b) Kiwifruit
c) Apricot
d) Almond

119. Which of the following is a dwarfing rootstock of plum?

a) Grand Ferrade
b) St. Julian
c) Siberian-C
d) Pixy

120. Among temperate fruits, pruning intensity is most severe in:

a) Apple
b) Pear
c) Peach
d) Plum

121. Botanically, the fruit of jackfruit is:

a) Berry
b) Drupe
c) Sorosis
d) Pome

122. Which of the following is highly shade tolerant fruit crop?

a) Carambola
b) Coconut
c) Citrus
d) Mango

123. Muscat flavour in grape is due to the presence of:

a) Methyl anthranilate
b) Methyl salicylate
c) Ethyl-2-methyl butyrate
d) Ethyl hexonate

124. 'Udhayam' variety of banana was released in 2005 from:

a) IIHR, Bangalore
b) BARC, Bombay
c) NRCB, Tiruchirapalli
d) TNAU, Coimbatore

125. Which of the following is a member of guava family?
 a) Sapota b) Jamun
 c) Custard apple d) Phalsa

126. Stone fruit formation is a disorder of:
 a) Apricot b) Apple
 c) Annona d) Peach

127. Degreening of citrus fruit is done by:
 a) CaC b) CH, Br
 c) CH2 = CH d) None of these

128. The virus free planting material can be produced by:
 a) Petiole c) Internode
 b) Midrib d) Meristern cultures

129. India has highest productivity of ....in the world.
 a) Mango b) Papaya
 c) Banana d) Grape

130. Skin freckling is occurs in:
 a) Citrus b) Apple
 c) Papaya d) Grape

131. Waxing helps in increasing storage life of fresh fruits by:
 a) Reducing evaporation
 b) Reducing respiration
 c) Reducing evaporation and respiration
 d) Arresting microbial growth

132. On the basis of ethylene production rate, passion fruit can be classified into:
 a) Low c) High
 b) Moderate d) Very high

133. Fruit which is cured in smoke for ripening:
 a) Apple b) Mango
 c) Banana d) Orange

134. Which of the following variety of banana is suitable for higher elevation?
 a) Ney Poovan b) Virupakshi
 c) Karupurvalli d) Kunnan

135. The structure of greenhouse which protects from wind is:
   a) Column b) Arches
   c) Purlin d) Bracing

136. Pruning intensity is depend on:
   a) Bearing habit b) Genotype
   c) Season d) Nutrients

137. Which of the following is practiced to obtain larger fruits?
   a) Training b) Pruning
   c) Thinning d) Chemical spray

138. Which of the following protected structure is suitable for hilly areas?
   a) Poly house b) Glass house
   c) Poly tunnel d) Lath house

139. In naturally ventilated greenhouse, temperature is maintained through:
   a) Vent b) Fuel
   c) Gas d) None of these

140. Which of the following is a dwarfing rootstock of mango?
   a) Rumani b) Creeping
   c) Moovadan d) Langra

141. B-box contains:
   a) DNA b) RNA
   c) Protein d) All of these

142. Stretching a piece of DNA is known as:
   a) Annotation b) Sequence alignment
   c) DNA combing d) DNA sequencing

143. Optimum pH for orchard planting is:
   a) 5.5-6.5 b) 6.5-7.5
   c) 7.5-8.5 d) 8.5-9.5

144. The main aim of crop regulation is:
   a) To force a tree for rest b) To induce off-season flowering
   c) To minimize pest infestation d) To check the growth

145. Gynoecious variety of papaya is:
   a) Pusa Giant b) Pusa Dwarf
   c) Sunrise Solo d) Pusa Nanha

Answers Key

1.	(b)	2.	(a)	3.	(c)	4.	(a)	5.	(c)	6.	(d)	7.	(b)
8.	(b)	9.	(b)	10.	(a)	11.	(d)	12.	(d)	13.	(c)	14.	(c)
15.	(d)	16.	(a)	17.	(b)	18.	(a)	19.	(b)	20.	(b)	21.	(b)
22.	(c)	23.	(b)	24.	(d)	25.	(c)	26.	(a)	27.	(c)	28.	(a)
29.	(b)	30.	(c)	31.	(d)	32.	(b)	33.	(c)	34.	(a)	35.	(a)
36.	(a)	37.	(b)	38.	(a)	39.	(d)	40.	(d)	41.	(a)	42.	(b)
43.	(b)	44.	(b)	45.	(d)	46.	(c)	47.	(c)	48.	(b)	49.	(c)
50.	(d)	51.	(b)	52.	(d)	53.	(b)	54.	(a)	55.	(c)	56.	(b)
57.	(c)	58.	(a)	59.	(b)	60.	(a)	61.	(c)	62.	(c)	63.	(c)
64.	(a)	65.	(a)	66.	(d)	67.	(c)	68.	(a)	69.	(b)	70.	(b)
71.	(c)	72.	(c)	73.	(a)	74.	(b)	75.	(d)	76.	(b)	77.	(c)
78.	(b)	79.	(c)	80.	(d)	81.	(c)	82.	(c)	83.	(a)	84.	(a)
85.	(c)	86.	(a)	87.	(c)	88.	(a)	89.	(a)	90.	(b)	91.	(c)
92.	(a)	93.	(c)	94.	(c)	95.	(c)	96.	(a)	97.	(d)	98.	(c)
99.	(a)	100.	(d)	101.	(a)	102.	(d)	103.	(b)	104.	(a)	105.	(d)
106.	(c)	107.	(d)	108.	(d)	109.	(b)	110.	(d)	111.	(d)	112.	(c)
113.	(b)	114.	(d)	115.	(d)	116.	(a)	117.	(c)	118.	(b)	119.	(d)
120.	(c)	121.	(c)	122.	(a)	123.	(b)	124.	(c)	125.	(b)	126.	(c)
127.	(c)	128.	(d)	129.	(d)	130.	(c)	131.	(c)	132.	(d)	133.	(c)
134.	(b)	135.	(d)	136.	(a)	137.	(c)	138.	(c)	139.	(a)	140.	(b)
141.	(c)	142.	(c)	143.	(b)	144.	(a)	145.	(c)				

14

ICAR – NET Fruit Science Exam – 2019

1. Which of the following variety of grape is commercially grown in North-India?
   a) Pearlette b) Pusa Seedless
   c) Thompson Seedless d) Pusa Navrang
2. Which of the following is highly shade tolerant fruit crop?
   a) Carambola b) Coconut
   c) Citrus d) Mango
3. Type of apomixis present in citrus is:
   a) Non recurrent b) Recurrent
   c) Parthenocarpy d) All of the above
4. Dwarf cultivar of peach is:
   a) Flordasun b) Early Grande
   c) Red Haven d) J. H. Hale
5. 'Nugget' is a dwarf cultivar of:
   a) Apple b) Pear
   c) Peach d) Plum
6. Training method followed in peach is:
   a) Central leader b) Open center
   c) Modified central leader d) Multiple system
7. Mango hybridization work was started in :
   a) 1901 b) 1911
   c) 1921 d) 1931
8. 'Mudkhed seedless is the variety of:
   a) Sweet orange b) Mandarin
   c) Grapefruit d) Lemon

9. Fruit cracking in pear is due to the deficiency of:
   a) Ca c) Mg
   b) B d) N
10. Grape variety suitable for double cropping in South India is:
   a) Arkavati b) Arka Hans
   c) Arka Kanchan d) Arka Shyam
11. Ultra dwarf rootstock of apple is:
   a) M-9 c) MM-106
   b) M-27 d) MM-104
12. Genomic constitution of FHIA-1 (Gold finger) is:
   a) AAAA c) AAAB
   b) ABBB d) ABB
13. Training system followed in cherry is:
   a) Open centre b) Central leader system
   c) Modified leader system d) None of the above
14. 'Rudrakshi' is the variety of:
   a) Guava b) Jamun
   c) Kiwifruit d) Jackfruit
15. Which of the following is a non climacteric fruit ?
   a) Apple c) Citrus
   b) Banana d) Mango
16. Inflorescence of fig is known as:
   a) Hypanthodium b) Panicle
   c) Catkin d) Balusta
17. Propagation method of cherry is:
   a) Tongue grafting b) T-budding
   c) Hardwood cutting d) Air layering
18. Iron deficiency first occurs in:
   a) Older leaves b) Younger leaves
   c) Nodes d) Buds
19. Foundation pruning in grapes is done in the month of:
   a) April b) October
   c) September d) November

20. The 'Earth Summit' (Biodiversity Convention) in Rio de Janeiro was held in:
    b) 1992 a) 1991
    c) 1993 d) 1994
21. Vernalization is a/an:
    a) Anaerobic process b) Aerobic process
    c) Both d) No effect of aerobic & anaerobic
22. The centre of origin of jackfruit is:
    a) Western Ghats b) Himalayan region
    c) Indo - Burma region d) Sunda island
23. The centre of origin of mango is:
    a) Indo - Burma region b) Sunda island
    c) Florida d) South west Asia
24. The state with least area under drip irrigation is:
    a) Andhra Pradesh b) Maharashtra
    c) West Bengal d) Manipur
25. Water saving in sprinkler system as compared to traditional irrigation system is:
    a) 10-15% b) 35-50%
    c) 50-65% d) 70-85%
26. Total number of biodiversity hotspots in the world are:
    a) 20 b) 25
    c) 30 d) 35
27. The precooling of litchi is done at:
    a) -5 °C c) 10 °C
    b) 5 °C d) 15 °C
28. Which of the following is a rapid method of precooling?
    a) Air cooling b) Hydro cooling
    c) Forced air cooling d) Vacuum cooling
29. Optimum temperature for banana cultivation is:
    a) 25.5 °C b) 26.5 °C
    c) 27.5 °C d) 28.5 °C
30. The fruit crop which is highly susceptible to waterlogging is:
    a) Grape b) Banana
    c) Pineapple d) Papaya

31. TSS of RTS is:
    b) 20°B a) 10°B
    c) 30°B d) 40°B
32. FPO stands for:
    a) Fruit Product Order b) Food Process Order
    c) Food Product Order d) Fruit Protect Order
33. Virus free plants can be obtained through:
    a) Anther culture b) Meristem culture
    c) Embryo culture d) Ovule culture
34. Which of the following is not a greenhouse gas?
    a) CO_2 b) O_2
    c) SO_2 d) CH_4
35. The chilling requirement of apple is:
    a) 600-800 hrs. b) 1000-1500 hrs.
    c) 1500-2000 hrs. d) None of the above
36. The auto-tetraploid fruit crop is:
    a) Ber b) Mango
    c) Kiwifruit d) European plum
37. Bending operation is done in:
    a) Ber b) Litchi
    c) Mango d) Guava
38. Protogyny is found in which fruit crop?
    a) Annona b) Walnut
    c) Passion fruit d) Sapota
39. Mridula pomegranate is a hybrid of:
    a) Ganesh × Gulshared
    b) Ganesh × Nana
    c) Open pollinated F, of Ganesh x Gulshared
    d) Open pollinated F, of Ganesh x Nana
40. The status of Geographical Indication (GI) tag is not given to:
    a) Virupakshi Hill Banana b) Nagpur Mandarin
    c) Bangalore Blue Grapes d) Zardalu Mango

41. Krishna Bhog mango is indigenous to:
    a) Odisha c) Bihar
    b) West Bengal d) Andhra Pradesh
42. The leading state in pineapple production in India is:
    a) Maharashtra b) Assam
    c) West Bengal d) Tripura
43. Spacing followed in meadow orcharding is:
    a) 1 × 2 m b) 2 x 5 m
    c) 5 x 5 m d) 2 × 3 m
44. Removal of suckers is commonly practiced in:
    a) Chestnut c) Hazelnut
    b) Pecan nut d) Pistachio nut
45. Gibberellins are derivatives of:
    a) Monoterpenes b) Diterpenes
    c) Triterpenes d) Sesquiterpene
46. Spring budding is common in which fruit crop:
    a) Mango b) Citrus
    c) Mangosteen d) Guava
47. 'Pusa Nanha' variety of papaya is developed through:
    a) Ploidy manipulation b) Hybridization
    c) Mutation d) Clonal selection
48. Rambutan (*Nephelium lappaceum*) belongs to the family:
    a) Rutaceae b) Sapindaceae
    c) Myrtaceae d) Rubiaceae
49. Karonda (*Carissa carandas*) belongs to the family:
    a) Boraginaceae b) Tiliaceae
    c) Apocynaceae d) Capparidaceae
50. English walnut is originated from:
    a) Europe b) North America
    c) South America d) Persia
51. Which of the following rootstock is used to hasten the maturity of kinnow mandarin?
    a) Troyer citrange b) Karna khatta
    c) Soh Sarkar d) Jattikhatti

52. The parents of Tangelo citrus are:

a) Mandarin × Sweet orange b) Mandarin × Lemon
c) Mandarin × Grapefruit d) Mandarin × Pummelo

53. Chromosome number of seedless guava is:

a) 2n = 30 b) 2n = 33
c) 2n = 24 d) 2n = 28

54. Phototropism was discovered by:

a) Charles Darwin b) Garner and Allard
c) Chailakhyan d) Lysenko

55. Palmette system of training is a modification of:

a) Central leader b) Spindle bush
c) Dwarf pyramid d) Open center

56. 'Button hole' is a harvesting disorder of:

a) Citrus c) Banana
b) Mango d) Grape

57. Which of the following is a triploid variety of apple ?

a) Baldwin b) Red June
c) Fuzi d) Florina

58. Lincoln canopy method of training is developed in:

a) Australia b) Italy
d) New Zealand c) Germany

59. 'Quince-C' is a dwarfing rootstock of:

a) Apple c) Peach
b) Pear d) Plum

60. Bartlett is a/an __________ variety of pear.

a) Early season b) Mid-season
c) Late season d) Extra early season

61. Which of the following is the highest yielder training system of grapevines ?

a) Bower b) Head
c) Kniffin d) Trellis

62. Red colour of 'Delicious' apples are the result of:
 a) Sectorial chimera b) Mericlinal chimera
 c) Periclinal chimera d) Simple chimera
63. 'Ambred' is a cross of:
 a) Red Delicious × Ambri b) Golden Delicious × Ambri
 c) Ambri × Red Delicious d) Straking Delicious × Ambri
64. Nendran banana belongs to which genomic group?
 a) AAA b) AAB
 c) ABB d) AB
65. Which of the following is a dwarfing rootstock of mango?
 a) Vellai Colomban b) Langra
 c) Alphonso d) Rumani
66. Molecular markers such as RFLP/ AFLP/RAPD can be used for:
 a) Cultivar identification b) Genetic diversity
 c) Phylogenetic relationships d) All of these
67. Which of the following is a growth retardant?
 a) IBA b) Kinetin
 c) GA d) Cycocel
68. Which of the following is the immediate precursor of ethylene biosynthesis?
 a) Methionine b) SAM
 c) ACC d) None of these
69. Fruit ripening hormone is:
 a) Auxin b) ABA
 c) Cytokinin d) Ethylene
70. 'Fruit thinning hormone is:
 a) NAA b) Ethrel
 c) DNOC d) Gibberellins
71. The growth regulator commonly used as weedicide:
 a) NAA b) ABC
 c) 2,4-D d) IBA
72. The common growth regulator for grape berry thinning and elongation is:
 a) NAA b) Ethrel
 c) BA d) GA_3

73. Which of the following variety is a source of resistant to malformation?

a) Bhadauran
b) Bombay Green
c) Husnara
d) Kesar

74. Mango malformation can be overcome by application of:

a) NAA @100 ppm
b) NAA @200 ppm
c) 2,4-D @200 ppm
d) Paclobutrazol @200 ppm

75. Denavelling is done in:

a) Papaya
c) Apple
b) Banana
d) Pineapple

76. Botanically, the fruit of apple is:

a) Berry
b) Drupe
c) Sorosis
d) Pome

77. Synthetic seeds are:

a) Artificially encapsulated somatic embryos
b) Artificially encapsulated zygotic embryos
c) Artificially encapsulated somatic hybrids
d) Artificially encapsulated zygotic hybrids

78. In which fruit crop, tissue culture is commercially exploited?

a) Apple
b) Mango
c) Aonla
d) Banana

79. The term 'Appertizing' is used for:

a) Canning
b) Dehydration
c) Syruping
d) Sterilization

80. The main site of auxin synthesis is:

a) Leaves
b) Root meristem
c) Shoot apex
d) Fruits

81. Which of the following is used for vigour control in apple?

a) SADH
b) 2,4-D
c) ABA
d) MH

82. Which of the following is a mutant variety of mango?

a) Rumani
c) Mulgoa
b) Rosica
d) Sensation

83. Sapota is commercially propagated by:
    a) Budding c) Inarching
    b) Grafting d) Layering
84. Multistem training is practiced in:
    a) Phalsa b) Pomegranate
    c) Citrus d) All of these
85. Hypobaric storage is also known as:
    a) Low temperature storage b) Low pressure storage
    c) Cold storage d) CA storage
86. Which of the following amino acid is precursor of auxin synthesis?
    a) Methionine c) Tryptophan
    b) Lysine d) Leucine
87. Chemical nature of kinetin is:
    a) 6-Furfuryl-aminopurine b) 5-Furfuryl-amino purine
    c) 6-Furfuryl-aminopyrimidine d) 5-Furfuryl-aminopyrimidine
88. Dogridge is a salt tolerant rootstock of:
    a) Mango b) Guava
    c) Grape d) Citrus
89. The crop which does not belongs to the same family is:
    a) Mango b) Cashew
    c) Pistachio nut d) Strawberry
90. Browning reaction involves:
    a) Sugars and minerals b) Sugars and fat
    c) Sugars and amino acids d) None of these
91. Most suitable packaging material for grapes is:
    a) Ply-board boxes b) Plastic boxes
    c) Corrugated fibre board boxes d) Wooden boxes
92. The oxygen concentration at which anaerobic respiration commence is known as:
    a) Respiration point b) Extinction point
    c) Critical point d) Danger point
93. Softnose, a post-harvest disorder of mango is occurs due to:
    a) Convective heat b) Coal fumes of brick kilns
    c) Ca deficiency d) All of these

94. Fruit of epigynous flower, formed in spike:
  a) Apple b) Pineapple
  c) Rose apple d) Custard apple
95. Which of the following citrus rootstock is not used commercially?
  a) Tahiti lime b) Rough lemon
  c) Rangpur lime d) Jatti khatti
96. Pre- harvest fruit drop in citrus can be controlled by spray of:
  a) BA b) GA
  c) ABA d) 2,4-D
97. Deficiency of which of the following leads dwarfing?
  a) Auxin b) Cytokinin
  c) Gibberellins d) Ethylene
98. Best maturity indices for citrus is:
  a) TSS b) Acidity
  c) TSS: Acid ratio d) Juice content
99. Chromosome number of citrus is:
  a) 2n = 16 b) 2n = 18
  c) 2n = 24 d) 2n = 28
100. Training system followed in passion fruit is:
  a) T-bar b) Two arm kniffin
  c) Matted row d) Four arm kniffin
101. Pruning in apple is done during:
  a) Dormant season b) Summer season
  c) Spring season d) Rainy season
102. Which of the following is a variety of kiwifruit:
  a) Allison c) Fuertes
  b) Gala d) President
103. Hyperhydricity (previously known as vitrification) is a disorder in:
  a) In vitro b) Ex vitro
  c) Hardening d) All of these
104. Which of the following product is the result of bacterial fermentation?
  a) Vinegar b) Cider
  c) Fanny d) Perry

105. Leaf bronzing in guava is due to:

a) Zn deficiency
b) Fe toxicity
c) Al toxicity
d) All of these

106. Little leaf in mango is associated with the deficiency of:

a) Ca
b) Mn
c) Zn
d) Mg

107. 'Mosambi' a sweet orange cultivar was introduced from:

a) Mozambique
b) Brazil
c) Pakistan
d) Iran

108. Which type of banana bear only female flower bud?

a) French plantain
b) Horn plantain
c) French horn
d) False horn

109. In which of the following parthenocarpy is not seen?

a) Citrus
b) Mango
c) Banana
d) Pineapple

110. Cymose type of inflorescence is found in:

a) Mango
b) Guava
c) Pear
d) Banana

111. Internal fruit necrosis of aonla can be controlled by spray of:

a) Borax 0.6%
b) Calcium chloride 0.2%
c) Mencozeb 0.6%
d) Sevin 0.3%

112. The fruit crop having basic chromosome number 7, but octaploid in nature:

a) Strawberry
b) Raspberry
c) Aonla
d) Blueberry

113. Which refrigerant is commonly used in cold storage in our country?

a) Ethylene
b) Carbide
c) Ammonia
d) Liquid nitrogen

114. Shoot tip grafting has been standardized for production of virus free plants in:

a) Banana
c) Mango
b) Citrus
d) Apple

115. Micro-budding is performed:

a) To produce virus free plants
b) To overcome incompatibility
c) To reduce the juvenile phase
d) All of these

116. Bananas turn black in the refrigerator due to:
    a) Freezing injury
    b) Chilling injury
    c) Both
    d) None of these

117. Epicotyle grafting is commercially followed in:
    a) Citrus
    b) Apple
    c) Mango
    d) Jackfruit

118. Mostly used protected structure by nurserymen/propagators in India is:
    a) Plastic tunnel
    b) Poly house
    c) Net house
    d) Glass house

119. 1-MCP is a:
    a) Ripening enhancer
    b) Ripening inhibitors
    c) Growth retardant
    d) None of these

120. Plants containing nucleus of one species but cytoplasm from both the parental species are known as:
    a) Somatic hybrids
    b) Cybrids
    c) Hybrids
    d) Haploids

121. Gutter height of a greenhouse is:
    a) 2-2.5 meter
    b) 3-3.5 meter
    c) 4-4.5 meter
    d) 5-5.5 meter

122. The thickness of UV stabilized plastic film used as glazing material for greenhouse is:
    a) 100 micron
    b) 200 micron
    c) 300 micron
    d) 400 micron

123. The thickness of LDPE (Low-density polyethylene) film used for mulching of banana is:
    a) 100 micron
    b) 200 micron
    c) 300 micron
    d) 400 micron

124. Protected structure with precise environmental controls is known as:
    a) Greenhouse
    b) Glasshouse
    c) Phytotron
    d) Polytunnel

125. The recommended storage temperature to extend the shelf-life of banana is:
    a) 2-3 °C
    b) 3-4 °C
    c) 27-29 °C
    d) None of these

126. The term 'Clinching' is associated with:
   a) Drying b) Canning
   c) Pickling d) Picking
127. The primary hormone associated with abscission:
   a) Auxin b) Cytokinin
   c) ABA d) Ethylene
128. Which of the following is not a gibberellin inhibitor?
   a) CCC b) AMO
   c) BA d) Paclobutrazol
129. Double sigmoid growth curve is observed in:
   a) Mango b) Apple
   c) Grape d) Citrus
130. Which of the following species of guava is a shrub?
   a) *Psidium cattleianum* b) *Psidium montanum*
   c) *Psidium molle* d) *Psidium pyriferum*
131. Browning of apple is due to:
   a) Malondialdehyde b) Carotenoids
   c) Anthocyanins d) All of these
132. Total number of endangered plant species in India are:
   a) 3364 b) 4464
   c) 5564 d) 2264
133. The incompatibility in fruit crops grafting can be overcome by:
   a) Bridge grafting b) Veneer grafting
   c) Epicotyl grafting d) Side grafting
134. Which of the following is the highest salt tolerant fruit crop?
   a) Date palm b) Ber
   c) Aonla d) Guava
135. Which of the following can be grown in pH level of 9-12?
   a) Mango b) Citrus
   c) Pomegranate d) Aonla
136. The formation of vertical water sprouts from horizontal branches is due to:
   a) Phototropism b) Geotropism
   c) Photomorphism d) None of these
137. Apple is trained through:
   a) Central leader system b) Open centre system
   c) Modified central leader system d) Trellis system

138. Most commonly followed training system in high density planting of apple is:

a) Spindle bush
c) Cordon
b) Tatura trellis
d) Palmette

139. Training system in which trees are more conical shaped with a distinct, supported, vertical central leader is known as:

a) Slender spindle
b) Cordon
c) Central leader
d) Vertical axis

140. Rapid and mass multiplication of true to type plants using tissue culture is known as:

a) Clonal propagation
b) Micro-propagation
c) Vegetative propagation
d) None of these

141. Sublimation is the principle behind:

a) Vacuum drying
b) Freeze drying
c) Spray drying
d) Flash drying

142. Which of the following is/are limitation (s) of drip irrigation?

a) Sensitivity to clogging
b) Moisture distribution
c) Salinity hazards
d) All of these

143. Which of the following is targeted for delaying ripening in genetic engineering of fruit crops?

a) CCC
b) ACC
c) SAM
d) All of these

144. To increase the humidity and reduce the e temperature in greenhouse, which of the following is practiced?

a) Sprinkler irrigation
b) Drip irrigation
c) Fogging
d) Flooding

145. The equilibrium relative humidity of absolute distilled water is:

a) 50%
b) 75%
c) 98%
d) 100%

146. Best rootstock of citrus in south India is:

a) Cleopatra mandarin
b) Citrange
c) Rangpur lime
d) None of these

147. Mango leaves shows drying scorching effect due to deficiency of:

a) N
b) Fe
c) K
d) Ca

148. Which of the following citrus rootstock is not propagated by seeds?

a) Tahiti lime b) Rough lemon

c) Rangpur lime d) Trifoliate orange

149. Which part of the plant is rich in Phytochrome?

a) Etiolated seedlings b) Light grown seedlings

c) Stems d) Leaves

150. Seed dormancy can be overcome by:

a) Stratification b) Scarification

c) Thiourea d) All of these

Answers

1.	(a)	2.	(a)	3.	(a)	4.	(c)	5.	(a)	6.	(b)	7.	(b)
8.	(b)	9.	(a)	10.	(d)	11.	(b)	12.	(c)	13.	(c)	14.	(d)
15.	(c)	16.	(a)	17.	(a)	18.	(b)	19.	(a)	20.	(b)	21.	(b)
22.	(a)	23.	(a)	24.	(d)	25.	(b)	26.	(d)	27.	(b)	28.	(d)
29.	(b)	30.	(d)	31.	(a)	32.	(a)	33.	(b)	34.	(b)	35.	(b)
36.	(a)	37.	(d)	38.	(a)	39.	(c)	40.	(d)	41.	(b)	42.	(c)
43.	(a)	44.	(c)	45.	(b)	46.	(b)	47.	(c)	48.	(b)	49.	(c)
50.	(d)	51.	(a)	52.	(c)	53.	(b)	54.	(a)	55.	(a)	56.	(a)
57.	(a)	58.	(d)	59.	(b)	60.	(b)	61.	(a)	62.	(c)	63.	(a)
64.	(b)	65.	(a)	66.	(d)	67.	(d)	68.	(c)	69.	(d)	70.	(a)
71.	(c)	72.	(d)	73.	(a)	74.	(b)	75.	(b)	76.	(d)	77.	(a)
78.	(d)	79.	(a)	80.	(c)	81.	(a)	82.	(b)	83.	(c)	84.	(b)
85.	(b)	86.	(c)	87.	(a)	88.	(c)	89.	(d)	90.	(c)	91.	(c)
92.	(b)	93.	(c)	94.	(b)	95.	(a)	96.	(d)	97.	(c)	98.	(c)
99.	(b)	100.	(b)	101.	(a)	102.	(a)	103.	(a)	104.	(a)	105.	(a)
106.	(c)	107.	(a)	108.	(b)	109.	(b)	110.	(b)	111.	(a)	112.	(a)
113.	(c)	114.	(b)	115.	(a)	116.	(b)	117.	(c)	118.	(c)	119.	(b)
120.	(b)	121.	(d)	122.	(b)	123.	(b)	124.	(c)	125.	(d)	126.	(b)
127.	(d)	128.	(c)	129.	(c)	130.	(b)	131.	(a)	132.	(c)	133.	(a)
134.	(a)	135.	(d)	136.	(a)	137.	(c)	138.	(a)	139.	(a)	140.	(b)
141.	(b)	142.	(d)	143.	(b)	144.	(a)	145.	(d)	146.	(c)	147.	(c)
148.	(a)	149.	(a)	150.	(d)								

15

ICAR – NET Fruit Science Exam – 2020

1. Colt, a dwarfing rootstock of cherry is a cross between:
   a) *P. avium* × *P. cerasifera* b) *P. avium* × *P. mahaleb*
   c) *P. avium* × *P. pseudocerasus* d) *P. avium* × *P. cerasoides*
2. GF-677 is a cross of:
   a) Almond × Peach b) Peach × Almond
   c) Peach × Apricot d) Apricot × Peach
3. Coat protein mediated resistance has been developed in:
   a) Papaya b) Banana
   c) Citrus d) All the above
4. Fleshy pericarp is the edible part of:
   a) Custard apple b) Apple
   c) Fig d) Bael
5. Term 'Ostiole' is associated with:
   a) Fig b) Persimmon
   c) Olive d) None of these
6. 'Attacin E' gene is used to impart resistance against:
   a) Fungus c) Bacteria
   b) Virus d) Viroids
7. M 27 rootstock was evolved from the cross:
   a) M13 × M7 b) M13 × M9
   c) M7 × M13 d) M9 × M13
8. Ultra dwarf rootstock of apple is:
   a) M 9 b) M 27
   c) P 22 d) MM 104
9. Which of the following is an ultra-dwarfing rootstock for citrus:
   a) Troyer citrange b) Rangpur Lime
   c) Flying Dragon d) Sour orange

10. 'Barcelona' is a famous variety of:
    a) Filbert b) Pistachio nut
    c) Chestnut d) Brazil nut
11. Type of incompatibility found in apricot is:
    a) Sporophytic b) Gametophytic
    c) Both d) Cross
12. Mixed bud is seen in:
    a) Apricot b) Plum
    c) Cherry d) Pear
13. Tatura trellis system of training was developed in:
    a) USA b) Germany
    c) New Zealand d) Australia
14. Ecodormancy is due to:
    a) Extreme temperature b) Physiological factors
    c) Hormonal imbalance d) Apical dominance
15. Which of the following is a transgenic variety of papaya ?
    a) Sinta b) Hortus Gold
    c) Red Lady d) Rainbow
16. Which of the following is/ are low chilling peaches?
    a) Sharbati b) Shan e Punjab
    c) Florda Sun d) All of these
17. The term 'Appertizing' is used for:
    a) Canning b) Dehydration
    c) Syruping d) Sterilization
18. Synersis of jelly is occurred due to:
    a) Excess of acid b) Overcooking
    c) Excess of pectin d) Lack of pectin
19. Sugar acts as a preservative by:
    a) Antioxidant b) Osmosis
    c) Inactivating enzymes d) Creating anaerobic condition
20. Which of the following enzyme is responsible for conversion of pectin into pectic acid?
    a) Protopectinase b) Pectic methyl esterase (PME)
    c) Pectinase d) Polygalcturonase

21. The 'Fluid Mosaic Model' of cell membranes was given by:
   a) Singer and Nicolson b) Davidson and Danielli
   c) David Robertson d) Ernest Overton

22. 'Nicking' is the partial ringing of branches:
   a) Above a dormant bud b) Below a dormant bud
   c) Both d) None of these

23. Storage temperature of mango is:
   a) 9-10°C b) 12-13°C
   c) 14-15°C d) 7-8°C

24. In banana, balbisiana genome is used to impart tolerance against:
   a) Flood c) Cold
   b) Drought d) Heat

25. The concentration of benzoic acid for preservation of squash and cordial is:
   a) 100 ppm b) 350 ppm
   c) 600 ppm d) 700 ppm

26. Potassium metabisulfite is chemically known as:
   a) $K_2S_2O_5$ b) $K_0S_2O_5$
   c) KSO_3 d) KSO_{11}

27. Benzoic acid is effective against:
   a) Fungi b) Bacteria
   c) Yeast d) Mould

28. Most important post-harvest factor responsible for deterioration of fruits is:
   a) Transpiration b) Respiration
   c) Ethylene evolution d) None of these

29. Who is the present chairman of IPCC (Intergovernmental Panel on Climate Change)?
   a) Rajendra K. Pachauri b) Hoesung Lee
   c) Robert Watson d) Bert Bolin

30. In Tamil Nadu, 'winter pruning' in grapevines is practiced during:
   a) October b) November
   c) December d) January

31. 'Paradox' is a rootstock of:
    a) Apricot b) Almond
    c) Cherry d) Walnut
32. 'Skin freckling' is a disorder of:
    a) Avocado b) Persimmon
    c) Papaya d) Apple
33. Mango rootstock 'Latara' is:
    a) Highly dwarfing b) Moderate dwarfing
    c) Moderate vigorous d) Highly vigorous
34. Micro-propagation technique is most successful in propagation of:
    a) Guava c) Aonla
    b) Banana d) Mango
35. Which growth regulator is used as a substitute for low chilling treatment ?
    a) Ethrel b) 2,4-D
    c) Kinetin d) Gibberellin
36. Most stable sex form in papaya is:
    a) Male b) Female
    c) Hermaphrodite d) All of these
37. Haploid plants are produced through:
    a) Pollen culture b) Ovule culture
    c) Meristem culture d) Cell culture
38. Which of the following grape variety sets fruits by stimulative parthenocarpy?
    a) Beauty Seedless b) Black Corianth
    c) Perlette d) Black Muscat
39. Which of the following is an orinthophillus fruit crop?
    a) Pineapple b) Banana
    c) Strawberry d) Oil palm
40. Which of the following shows very high ethylene production?
    a) Mango b) Passion fruit
    c) Grape d) Sapota
41. The hybrid formed by fusion of two somatic cells is known as:
    a) Cybrid b) Gametic hybrid
    c) Somatic hybrid d) Distant hybrid

42. 'Saigon' is a seedling rootstock of:

a)	Mango	c)	Citrus
b)	Cherry	d)	Apple

43. 'Notching' is practiced in:

a)	Fig	b)	Wood apple
c)	Jamun	d)	Grape

44. 'PDSD' is a unique feature of:

a)	Chestnut	b)	Pecan nut
c)	Custard apple	d)	Avocado

45. In north India, pruning in phalsa is done in:

a)	December - January	b)	March-April
c)	June - July	d)	September - October

46. Kiwifruit is a/an:

a)	Deciduous vine	b)	Evergreen vine
c)	Deciduous shrub	d)	Evergreen shrub

47. The basic chromosome number (X=) of jackfruit is:

a)	12	b)	14
c)	24	d)	28

48. MM series of apple rootstocks are resistant against:

a)	Virus	b)	Fire blight
c)	Woolly apple aphid	d)	Sanjose scale

49. Which of the following is a deciduous species of citrus?

a)	Fortunella hindsi	b)	Citrus unshui
c)	Poncirus trifoliata	d)	Microcitrus autraliaca

50. The inflorescence of ber is known as:

a)	Solitary	b)	Panicle
c)	Catkin	d)	Fasicle

51. Sapota grows mostly on:

b)	Spurs	a)	One year old shoots
c)	Current season growth	d)	Very old shoots

52. Flow cytometery is used for:

a)	Identification of polyploids	b)	Detection of virus
c)	Identification of hybrids	d)	Confirmation of mutants

53. The bud wood of temperate plant species is stored at:

a) 32°F c) 72°F

b) 48°F d) 96°F

54. The induction of flowering by exposure to the chilling treatment is known as:

a) Stratification b) Scarification

c) Vernalization d) Photoperiodism

55. Musca domestica is a principal pollinator of:

a) Fig b) Custard apple

c) Persimmon d) Mango

56. The undifferentiated mass of cells is known as:

a) Callus c) Media

b) Explant d) None of these

57. Aroma of over ripe banana is due to:

a) Diallyl propyl b) Hexanol

c) Allyl propyl d) Isopentanol

58. The conservation of fruit crops germplasm outside their natural habitat is known as:

a) Ex-situ conservation b) In-situ conservation

c) On farm conservation d) None of the above

59. Maximum concentration of genus Mangifera is found in:

a) Indian peninsula b) Malaya peninsula

c) Arabian peninsula d) North America

60. Dehorning is practiced in the high density planting of:

a) Grapes b) Apple

c) Peach d) Mango

61. Hedge row planting system is commercially followed in cultivation of:

a) Grape and guava b) Guava and pineapple

c) Apple and pineapple d) Apple and peaches

62. A type of culture in which small aggregates of cells multiply while suspended in agitated liquid medium is known as:

a) Single cell culture b) Suspension culture

c) Callus culture d) Organ culture

63. In aonla, the male flowers first appear in cluster in the ________ of leaves on lower part of the determinate shoot.

a) Basal
b) Tip
c) Axil
d) All of these

64. Gamma irradiation dose is sufficient to provide a high level of quarantine security against fruit fly in mango:

a) 100 Gy
c) 300 Gy
b) 200 Gy
d) 400 Gy

65. Which of the following is a dwarfing rootstock of mango?

a) Vellai Colomban
b) Langra
c) Alphonso
d) Rumani

Answers

1.	(c)	2.	(b)	3.	(d)	4.	(a)	5.	(a)	6.	(c)	7.	(b)
8.	(b)	9.	(c)	10.	(a)	11.	(b)	12.	(d)	13.	(d)	14.	(a)
15.	(d)	16.	(d)	17.	(a)	18.	(a)	19.	(b)	20.	(b)	21.	(a)
22.	(b)	23.	(b)	24.	(b)	25.	(c)	26.	(a)	27.	(c)	28.	(b)
29.	(b)	30.	(a)	31.	(d)	32.	(c)	33.	(a)	34.	(b)	35.	(d)
36.	(b)	37.	(a)	38.	(b)	39.	(a)	40.	(b)	41.	(c)	42.	(a)
43.	(a)	44.	(d)	45.	(a)	46.	(a)	47.	(b)	48.	(c)	49.	(c)
50.	(d)	51.	(c)	52.	(a)	53.	(a)	54.	(c)	55.	(d)	56.	(a)
57.	(d)	58.	(a)	59.	(b)	60.	(d)	61.	(c)	62.	(b)	63.	(c)
64.	(a)	65.	(a)										

16

ICAR – NET
Fruit Science Exam – 2021

1. 'Flying Dragon' a dwarfing rootstock of citrus. The dwarfness is governed by:
   a) Single dominant gene b) Single recessive gene
   c) Polygenes d) Additive genes
2. India's contribution in the total world production of fruits is.
   a) 10.5% b) 12.5%
   d) 17.5% c) 15.5%
3. In which fruit crop the first transgenic was produced:
   a) Papaya c) Pear
   b) Plum d) Walnut
4. International Biodiversity Day is celebrated on:
   a) 23 November b) 22 May
   c) 22 December d) 23 April
5. 'Manjri Naveen' variety of grape is a clonal selection from:
   a) Sharad Seedless b) Fantasy Seedless
   c) Crimson Seedless d) Centennial Seedless
6. The causal organism of mango malformation is:
   a) Bacteria b) Virus
   c) Fungus d) Nematode
7. 'Arka Kiran' is a hybrid variety of:
   a) Mango c) Papaya
   b) Guava d) Pomegranate
8. 'Pusa Srijan' is a dwarfing rootstock of:
   a) Citrus b) Guava
   c) Sapota d) Mango

9. Protogynous diurnal synchronous dichogamous flowering pattern occur in:
   a) Annona muricata b) Castenea sativa
   c) Pistacia vera d) Persea americana
10. High C: N ratio favours:
   a) Vegetative growth b) Reproductive growth
   c) Bud dormancy d) All of these
11. Seedlessness in most of the grape variety is due to:
   a) Vegetative parthenocarpy b) Stimulative parthenocarpy
   c) Stanospermocarpy d) None of these
12. Degreening is practiced in:
   a) Citrus b) Litchi
   c) Carambola d) Apple
13. "Top Red' is the bud sport of......... variety.
   a) Starking Delicious b) Red Delicious
   c) Shot Well Delicious d) Golden Delicious
14. Gamboge is a disorder in which of the following fruit:
   a) Avocado b) Mango
   c) Mangosteen d) Apricot
15. Which of the following fruit is rich in pectin ?
   a) Apple b) Guava
   c) Litchi d) Papaya
16. Fascination is a disorder of:
   a) Strawberry b) Banana
   c) Papaya d) Pineapple
17. Most widely grown cultivar of apple in the world is:
   a) Red Delicious b) Fuzi
   c) Gala d) Golden Delicious
18. 'Chaubatia Alnkar' and 'Chaubatia Madhu' are the varieties of:
   a) Apple c) Almond
   b) Peach d) Apricot
19. The somatic chromosome number of pecan nut is:
   a) 2n = 32 b) 2n=34
   c) 2n = 36 d) 2n = 38

20. Dashehari-51, mango variety a developed through:

a) Chance seedling selection b) Clonal selection

c) Hybridization d) Mutation

21. The sparkling clear liquid free from all suspended solid is called as:

b) Squash a) Concentrate

d) Beverage c) Cordial

22. Modified atmospheric packaging (MAP) of fruits prevent the building up of the following:

a) Protein b) Sugar

c) CO_2 and acetylene d) None of these

23. The recommended storage temperature to extend the shelf-life of banana is:

a) 4-5°C b) 8-9°C

c) 12-13°C d) 15-16°C

24. 'Bhagwa' is an open pollinated selection of:

a) Litchi b) Guava

c) Pomegranate d) Mangosteen

25. Training system followed in Kiwifruit is:

a) T-bar system b) Kniffin system

c) Spindle bush system d) Bower system

26. Spacing followed in meadow orcharding of guava is:

a) 1×2 m b) 2 × 5 m

c) 5×5 m d) 2 × 3 m

27. Photosynthetically active radiation (PAR) is measured by:

a) Pyranometer b) Pyrano - albedometer

c) Quantum sensor d) All of these

28. Photosynthetically active radiation (PAR) wave length range is:

a) 100 to 300 nm b) 400 to 700 nm

c) 700 to 1000 nm d) 1000 to 1300 nm

29. Removal of terminal portion of the shoots leaving its basal portion is known as:

a) Thinning out b) Heading back

c) Skirting d) Dehorning

30. The term 'Propping' is related to:
    a) Strawberry b) Banana
    c) Pineapple d) Papaya
31. Leading state in protected cultivation in India is:
    a) Himachal Pradesh b) Gujarat
    c) Maharashtra d) Andhra Pradesh
32. Artificial hybridization in fruit crop was first practiced by:
    a) Thomas Fairchild b) Luther Burbank
    c) Thomas Andrew Knight d) Camoreous
33. Which of the following is the nodal agency for introducing plant genetic resources in country?
    a) ICAR c) IARI
    b) NBPGR d) FRI
34. The cultivated grape (Vitis vinifera) is considered to be a hybrid of:
    a) *Vitis vulpine* × *Vitis labrusca* b) *Vitis vulpine* × *Vitis champini*
    c) *Vitis labrusca* × *Vitis rupesteris* d) *Vitis aestavilis* × *Vitis labrusca*
35. Parthenocarpy in fruits is induced by:
    a) GA b) Ethylene
    d) 2,4-D c) Zeatin
36. Which growth regulator is used as a substitute for low chilling treatment?
    a) Ethrel b) 2,4-D
    c) Kinetin d) Gibberellin
37. Which of the following is an ethylene absorbent?
    a) $KMnO_4$ c) KNO
    b) KCl_3 d) K_2SO_4
38. Which of the following is used for berry elongation in grapes?
    a) NAA c) Dormex
    b) Kinetin d) GA_3
39. 'D-leaf' is the best indicator of nutrient status of:
    a) Pineapple c) Banana
    b) Grape d) Apple
40. Most commonly used training system in apple is:
    a) Spindle bush b) Open center
    c) Modified central leader d) None of these

41. The undifferentiated mass of cells known as:
    a) Callus b) Explant
    c) Media d) None of these
42. The ability of a plant cell to develop into whole plant is known as:
    a) Prepotency b) Multipotency
    c) Totipotency d) Pluropotency
43. Cavendish banana belongs to the group:
    a) AAA b) AAB
    c) ABB d) AB
44. Kagzi lime (*Citrus aurantifolia*) is an indicator plant of:
    a) Citrus greening b) Citrus tristeza virus
    c) Citrus psorosis d) Citrus exocortis
45. The ideal percentage of fruit part and TSS in jelly should be:
    a) 45% fruit part and 55% TSS b) 45% fruit part and 65% TSS
    c) 55% fruit part and 45% TSS d) 65% fruit part and 45% TSS
46. The concept of 'Photoperiodism' was given by:
    a) Lysenko b) Garner and Allard
    c) Kidd and West d) F. W. Went
47. 'Hass' is a variety of:
    a) Avocado b) Strawberry
    c) Persimmon d) Carambola
48. 'Cork spot' is a disorder of:
    a) Peach c) Pear
    b) Plum d) Cherry
49. PKM-1 variety of acid lime is developed through:
    a) Selection b) Hybridization
    c) Mutation d) Polyploidy breeding
50. Genomic constitution of 'Gold Finger' (FHIA-1) is
    a) BBB b) AAAB
    c) ABBB d) AAAA
51. National Biodiversity Authority (NBA) headquartered in:
    a) New Delhi c) Hyderabad
    b) Chennai d) Bengaluru

52. Spacing followed in HDP of Amrapali variety of mango is:
    a) 2.5 x 2.5 m c) 2.5 x 5.5 m
    b) 3 x 2.5 m d) 5.5 x 5.5 m

53. Which of the following is viviparous in nature?
    a) Papaya c) Sapota
    b) Jackfruit d) Loquat

54. Anther or pollen culture is used for production of:
    a) Haploid plants b) Seedless plants
    c) Virus free plants d) None of these

55. Pre-harvest fruit drop in mango can be controlled by spraying:
    a) NAA c) GA
    b) 2,4-D d) Ethrel

56. Which of the following is a tenturier variety of grape?
    a) Pusa Uravshi b) Pusa Navrang
    c) Pusa Trishar d) Pusa Aditi

57. Stylar end is related to:
    a) Mango c) Litchi
    b) Mandarin d) Banana

58. A major challenge in the protected cultivation of fruit crops is the increasing threat of :
    a) White flies b) Aphids
    c) Fruit fly d) Nematodes

59. The major acid present in grape is:
    a) Citric acid b) Tartaric acid
    c) Malic acid d) Oxalic acid

60. Persimmon originated from:
    a) China b) Japan
    c) South America d) Tropical Africa

61. The somatic chromosome number of phalsa is :
    a) 2n = 26 c) 2n=32
    b) 2n=28 d) 2n=36

62. Bael is native of:
    a) Brazil b) India
    c) China d) Africa

63. Which of the following strawberry variety is known to be day neutral?
    a) Toiga c) Selva
    b) Sweet Charlie d) Pajaro

64. Process of removal of air from filled can is termed as:
    a) Blanching b) Exhausting
    c) Processing d) Sealing

65. Geographical Indication (GI) tag is valid for:
    a) 8 year b) 10 year
    c) 15 year d) 18 year

66. Important constituent of jelly is:
    a) Sugar b) Acid
    c) Water d) Pectin

67. 'Pusa Delicious' and 'Pusa Majesty' are the .... varieties of papaya.
    a) Dioeciuos b) Monoecious
    c) Gynodioecious d) Gynomonoecious

68. Which of the following fruit is parthenocarpic?
    a) Mango b) Citrus
    c) Banana d) Grape

69. Severe pruning is practiced in:
    a) Guava b) Ber
    c) Loquat d) Pomegranate

70. Which of the following is known as 'Star fruit'?
    a) Phalsa c) Kiwi
    b) Carambola d) Lasoda

71. Seed viability lost when the moisture content drops down to critical level in:
    a) Orthodox seeds b) Recalcitrant seeds
    c) Truthful label seeds d) None of these

72. The genetic variability present among the tissue cultured cells, plants is called as:
    a) Somaclonal variation b) Mutation
    c) Chimeras d) None of these

73. Markers used for determining genetic variation in plants is:
    a) Morphological markers b) Biochemical markers
    c) Molecular markers d) All of these

74. Haploids are used for:
    a) Development of homozygous lines
    b) Generation of exclusive male plants
    c) Induction of mutations
    d) All the above

75. Tissue cultured shoots of banana are stored in liquid nitrogen at:
    a) 0°C b) -206°C
    c) -196°C d) -60°C

76. Botanically, the fruit of apple is:
    a) Berry b) Drupe
    c) Sorosis d) Pome

77. Protected structure covered with plastic films is known as:
    a) Net house b) Shade house
    c) Poly house d) Glass house

78. M. Hakkinen (2013) reappraised the sectional taxonomy in Musa and reduced the number of sections in the genus Musa from 5 to:
    a) 1 b) 2
    c) 3 d) 4

79. The growth arresting PGR is:
    a) GA_3 b) NAA
    c) Zeatin d) CCC

80. In which fruit crop, fruit buds borne laterally with leafy shoots in the leaf axils ?
    a) Walnut b) Avocado
    c) Citrus d) Guava

81. CEPA is chemically:
    a) 2-Chloroethyl phosphoric acid b) 2-Chloromthyl phosphoric acid
    c) 2-Chloroethyl phosphonic acid d) 2-Chloromthyl phosphonic acid

82. Final stage of fruit development:
    a) Maturation b) Ripening
    c) Senescence d) Horticultural maturity

83. Lyophilization is also known as:
    a) Drying b) Freeze drying
    c) Freezing d) Dehydration

84. The process of breakdown of the protein in food (usually fatty foods) is known as:

a) Putrefaction b) Rancidity

c) Souring d) Sliminess

85. The main site of auxin synthesis is:

a) Leaves b) Root meristem

c) Shoot apex d) Fruits

86. Wine grape is botanically known as:

a) Vitis vinifera b) Vitis champini

c) Vitis labrusca d) Vitis aestavilis

87. Which of the following is a polygamous fruit crop?

a) Guava b) Avocado

c) Papaya d) Mandarins

88. The name 'Tamarind' derives from which language?

a) Arabic b) Latin

c) Greek d) Spanish

89. The crop which is grown in extremely hot regions?

a) Mango b) Guava

d) Aonla c) Date palm

90. Off season flowering is a peculiar phenomenon in mango occurs in which state?

a) Maharashtra b) Himachal Pradesh

c) Tamil Nadu d) Andhra Pradesh

91. In papaya, female plants can be produced in bulk through:

a) Pollen culture b) Ovule culture

c) Meristem culture d) Cell culture

92. Fire blight is a serious disease of:

a) Cherry b) Pear

c) Plum d) Peach

93. Forkert budding in mango is practiced in :

a) Brazil b) Venezuela

c) Indonesia d) Spain

94. Most important character of 'Filler tree' in fruit orchard is:

a) Quick growing
b) Spreading type
c) Slow growing
d) Late maturity

95. 'Chanee' is a variety of:

a) Durian
b) Dragon fruit
c) Hazel nut
d) Rambutan

96. Best time for application of paclobutrazol in mango is:

a) February-March
b) April - May
c) June-July
d) August-September

97. Which of the following rootstock is incompatible for most of the ber varieties?

a) *Ziziphus rotundifolia*
b) *Ziziphus mauritiana*
c) *Ziziphus jujube*
d) *Ziziphus nummularia*

98. Most devastating disease of almond is:

a) Bacterial gummosis
b) Collar rot
c) Brown rot
d) Downy mildew

99. Pruning in fruit crops is done to:

a) Increase yield
b) Obtain quality fruits
c) Prevent alternate bearing
d) None of these

100. In non-recurrent apomixis, embryo develops from:

a) Haploid cells
b) Diploid cells
c) Synergids cells
d) Antipodal cells

101. Zeatin is isolated from:

a) Sorghum
c) Mango
d) Carrot
b) Maize

102. PLR (J)-1 is a variety of:

a) Jamun
b) Jackfruit
c) Guava
d) Avocado

103. Indian Journal of Agricultural Sciences is published by:

a) IARI
c) IIHR
b) ICAR
d) NBPGR

104. Self-incompatibility in fruit crops is due to:
   a) Differences in floral morphology
   b) Physiological or genetic causes
   c) Environmental conditions
   d) All the above

105. Pink red colour in guava is controlled by:
   a) Dominant genes   b) Recessive genes
   c) Additive genes   d) Polygenes

106. During sealing of the can, temperature should not be fall below:
   a) 50°C   b) 40°C
   c) 64°C   d) 74°C

107. Which of the following amino acid is precursor of auxin synthesis?
   a) Methionine   b) Lysine
   c) Tryptophan   d) Leucine

108. Kew variety of pineapple belongs to the group:
   a) Queen   b) Spanish
   c) Cayenne   d) Abcaxi

109. Major limiting factor in grape cultivation under north Indian condition is :
   a) Winter frost   b) Early rain
   c) Drought   d) Salt

110. Andromonoecious type of inflore-scence is found in:
   a) Coconut   b) Arecanut
   c) Cashew nut   d) Oil palm

111. Which of the following method is used for genetic transformation of fruit crops?
   a) Electrofusion
   b) Agrobacterium mediated
   c) Microprojectile bombardment based
   d) Liposome mediated

112. Spacing between two walls of 'Zero Energy Cool Chamber'
   a) 5 cm   c) 10 cm
   b) 7.5 cm   d) 15 cm

113. Most commonly used explant for micro propagation of banana is:

a) Axillary buds
c) Shoot tip
b) Leaf sheath
d) Petiole

114. Gibberellin concentration used to increase the berry size of gapes:

a) 25-50 ppm
b) 100-200 ppm
c) 600-800 ppm
d) 800-1000 ppm

115. Avocado race having maximum size fruit is :

a) Mexican
b) Guatemalan
c) West Indian
d) None of these

116. Which of the following structure is suitable for hardening of rooted cuttings ?

a) Cold frame
b) Hot bed
c) Lath house
d) Net house

117. Preferred row orientation in fruit orchard is:

a) North-south
b) East - west
c) North-west
d) South-east

118. 'Nendran' banana belongs to which genomic group?

a) AAA
b) AAB
c) ABB
d) AB

119. How many basic stages are involved in fruit development?

a) 2
b) 3
c) 4
d) 5

120. Most appropriate method for storage of fruits is:

a) Cold storage
b) Refrigeration
c) Controlled atmospheric storage
d) Modified atmospheric storage

121. Gutter height of a greenhouse is:

a) 2-2.5 meter
b) 3-3.5 meter
c) 4-4.5 meter
d) 5-5.5 meter

122. Calculate the concentration of ethylene when 1ml dissolved in 1000ml of water:

a) 10 ppm
b) 100 ppm
d) 10000 ppm
c) 1000 ppm

123. In papaya, M_2 m express:

a) Male b) Sex reversing male
c) Hermaphrodite d) Female

124. Protected structure with precise environmental controls is known as:

a) Greenhouse b) Glasshouse
c) Phytotron d) Polytunnel

125. Macrophomina phaseolina is predomi nantly present in:

a) Saline soil b) Alkaline soil
c) Acidic soil d) Neutral soil

126. Pecan nut was introduced in India in:

a) 1668 c) 1735
b) 1879 d) 1935

127. Vavilov classified centre of origin in how many category:

a) 6 b) 7
c) 8 d) 12

128. For better canopy structure in fruit crops, the crotch angle should be:

a) 30° b) 45°
c) 60° d) 90°

129. Apple is commercially propagated by:

a) Tongue grafting b) Wedge grafting
c) Veneer grafting d) Softwood grafting

130. Which of the following structure is used for studying interactions between plants and the environment?

a) Glasshouse b) Polyhouse
c) Phytotron d) Lath house

131. Dichogamy is very much common in:

a) Citrus fruits b) Annonaceous fruits
c) Stone fruits d) Pome fruits

132. Which of the following species of papaya is tolerant against cold?

a) *Vasconcellea cundinamarcensis* b) *Vasconcellea stipulata*
c) *Vasconcellea quercifolia* d) *Vasconcellea goudotiana*

133. The incompatibility in fruit crops grafting can be overcome by:

a) Bridge grafting b) Veneer grafting
c) Epicotyl grafting d) Side grafting

134. Strawberries are harvested when .... of skin develops colour.

a) 25-50% b) 50-75%

c) 75-100% d) None of these

135. Which of the following berry is not grown in greenhouse?

a) Strawberry b) Raspberry

d) Blackberry c) Blueberry

136. Blueberries should be harvested at ?

a) Full colour development b) 4 colour development

c) 2/3rd colour development d) None of these

137. Chilling requirement of kiwifruit is:

a) 100-400 hrs. b) 400-700 hrs.

c) 700-1000 hrs. d) 1000-1200 hrs.

138. Most commonly used solution for osmotic dehydration:

a) Sugar solution b) KMS solution

c) Benzoic acid solution d) All the above

139. The ideal storage temperature for peach is:

a) 0-3°C b) 4-6°C

c) 5-8°C d) 9-12°C

140. Salt used as preservative have concentration of:

a) 2% b) 8%

c) 10% d) 15%

Answers Key

1.	(a)	2.	(b)	3.	(d)	4.	(b)	5.	(d)	6.	(c)	7.	(b)
8.	(b)	9.	(d)	10.	(b)	11.	(c)	12.	(a)	13.	(c)	14.	(c)
15.	(b)	16.	(d)	17.	(d)	18.	(d)	19.	(a)	20.	(b)	21.	(c)
22.	(c)	23.	(c)	24.	(c)	25.	(a)	26.	(a)	27.	(c)	28.	(b)
29.	(b)	30.	(b)	31.	(c)	32.	(c)	33.	(b)	34.	(a)	35.	(a)
36.	(d)	37.	(a)	38.	(d)	39.	(a)	40.	(c)	41.	(a)	42.	(c)
43.	(a)	44.	(b)	45.	(b)	46.	(b)	47.	(a)	48.	(c)	49.	(a)
50.	(b)	51.	(b)	52.	(a)	53.	(b)	54.	(a)	55.	(a)	56.	(b)
57.	(b)	58.	(d)	59.	(b)	60.	(a)	61.	(d)	62.	(b)	63.	(c)
64.	(b)	65.	(b)	66.	(d)	67.	(c)	68.	(c)	69.	(b)	70.	(b)
71.	(b)	72.	(a)	73.	(c)	74.	(d)	75.	(c)	76.	(d)	77.	(c)
78.	(b)	79.	(d)	80.	(b)	81.	(c)	82.	(a)	83.	(b)	84.	(a)
85.	(c)	86.	(a)	87.	(c)	88.	(a)	89.	(c)	90.	(c)	91.	(b)
92.	(b)	93.	(c)	94.	(a)	95.	(a)	96.	(d)	97.	(d)	98.	(a)
99.	(b)	100.	(a)	101.	(b)	102.	(b)	103.	(b)	104.	(b)	105.	(a)
106.	(d)	107.	(c)	108.	(c)	109.	(b)	110.	(c)	111.	(b)	112.	(b)
113.	(c)	114.	(a)	115.	(c)	116.	(c)	117.	(a)	118.	(b)	119.	(b)
120.	(c)	121.	(d)	122.	(c)	123.	(c)	124.	(c)	125.	(b)	126.	(d)
127.	(c)	128.	(b)	129.	(a)	130.	(c)	131.	(b)	132.	(a)	133.	(a)
134.	(b)	135.	(c)	136.	(a)	137.	(c)	138.	(a)	139.	(a)	140.	(d)

17

ICAR - NET Fruit Science Exam – 2023

1 In every winters peach requires heavy pruning because:
   a) It bears on current season's growth
   b) It bears on previous season's growth only
   c) It bears on two years' old wood only
   d) It bears on spur

2. Mango wilt/decline is spread through:
   a) Mealy bug   b) White grub
   c) Beetle   d) Leaf hopper

3. By using which standard rootstock of apple, ones can accommodate 4444 plants per hectare in an orchard:
   a) M-7   b) MM-104
   c) M-9   d) MM-111

4. Which of the following in not matched correctly:
   a) Date palm: monoecious   b) Papaya: polygamous
   (C) Walnut: Monoecious   d) Kiwifruit: dioecious

5. CO2 concentration in polyhouse should be:
   a) 0.11%   B) 0.20%
   C) 2.0%   d) 8.0%

6. *Mangifera dongnaiensis* and *M. orophila* cold species of mango are grown in the restricted area of:
   a) Mediterranean region   b) Central America
   c) Korean Island   d) Central Asia

7. Controlled environmental conditions having which two gases composition:
   a) CO_2 and NO_2 b) CO_2 and C_2H_4
   c) CO_2 and O_2 d) O_2 and N

8. Queen of the nut crops is:
   a) Apricot b) Walnut
   c) Pecan nut d) Pistachionut
9. *Musa veluntina* species of banana is native of which state of India:
   a) Assam b) Bihar
   c) Tripura d) Mizoram
10. Chaubattia Agrim variety of apple is developed through:
   a) Mutation b) Hybridization
   c) Clonal Selection d) Introduction
11. Papaya species require same environmental condition as *Carica papaya*, producing edible fruits and having resistant to PRSV:
   a) *V. cauliflora* b) *V. pubescence*
   c) *V. helbornii* d) *V. parviflora*
12. Which training system is most popular for establishing a HDP apple orchard:
   a) Spindle Bush b) Open vase
   c) Cordon d) Tatura trellis
13. Ploidy level of 'Case' variety of grape is:
   a) 3x b) 4x
   c) 6x d) 2x
14. Akbar variety of apple is a cross between:
   a) Ambri x Golden Delicious b) Ambri x Cox's Orange Pippin
   c) Starking Delicious x Ambri-81 d) Red Delicious x Ambri-57
15. In which banana variety only 6 leaves are minimum to require for flowering and normal fruit development without affecting fruit quality: (Source: NRCB website)
   a) Poovan b) Rasthali
   c) Karpuravalli d) Nendran
16. Chironji is commercially propagated by which method:
   a) Hard wood cutting b) Tongue grafting
   c) Softwood grafting d) Budding
17. In aonla ideal timing for performing patch budding is:
   a) February-March b) September – October
   c) June-July d) November – December

18. Widest spacing is followed in which fruit crop:
    a) Walnut b) Mango
    c) Jamun d) Chestnut
19. International Treaty on Plant Genetics Resources for Food and Agriculture (ITPGRFA) was first time signed in which year:
    a) 2001 b) 2003
    c) 2006 d) 2020
20. National Biodiversity Authority was established in:
    a) 2001 b) 2002
    c) 2003 d) 2004
21. Integrated FSSAI Act was enacted in:
    a) 1937 b) 2006
    c) 2008 d) 2011
22. Classification of banana by Simmonds and Shepherd given in which year:
    a) 1950 b) 1955
    c) 1957 d) 1958
23. In mango under the protected cultivation flowering can be induced by application of:
    a) PPP3 b) Ethrel
    c) MH d) NAA
24. Earliest variety of apricot is:
    a) Chaubattia Madhu b) Chaubattia Alankar
    c) Halman d) St. Ambroise
25. Cherry disease recently spreads in Himachal Pradesh is mainly caused by:
    a) Bacteria b) Phytoplasma
    c) Virus d) Fungus
26. Mango variety which did not get GI tag till date:
    a) Kishanbhog b) Malda
    c) Khirsapati d) Laxmanbhog
27. Mango good quality and sized fruits are grown in which system:
    a) Square system b) Rectangular system
    c) Double Hedge Row System d) Quincunx system

28. Which Annona species is having yellow pulp colour:
    a) *A. montana* b) *A. glabra*
    c) *Annona purpurea* d) *A. atemoya*
29. Fruit crop also known as "Heaven and Hell Fruit":
    a) Macadamia b) Durian
    c) Rambutan d) Jackfruit
30. North Star is a dwarf rootstock of which fruit crop:
    a) Peach b) Pear
    c) Plum d) Cherry
31. Draft genome in mango by Indian Genomic Consorting has been submitted for which of the following variety:
    a) Alphonso b) Amrapali
    c) Dashehari d) Rumani
32. Which of the following mango variety is a sister seedling of Pusa Peetamber:
    a) Pusa Shrestha b) Pusa Manohari
    c) Pusa Lalima d) Pusa Arunima
33. In Utah chilling model maximum chilling period are obtained in range of:
    a) 0-4°C b) 4-6°C
    c) 6-7°C d) 7-10°C
34. Botanical name of yellow skin dragon fruit species is:
    a) *Hylocereus undatus* b) *Hylocereus megalanthus*
    c) *Hylocereus costarisensis* d) *Hylocereus polyrhizus*
35. Gene editing for a fungal disease was first time done in which fruit:
    a) Cherry b) Peach
    c) Walnut d) Plum
36. After harvesting to remove the field heat from produce, most commonly precooling method followed is:
    a) Conduction b) Convention
    c) Vacuum cooling d) Dipping in water
37. Who develop plum pox virus resistant variety:
    a) R. Scorza b) H P Olmo
    c) Abhay Dandekar d) R Schomit

38. Apple variety "Arctic apple" was developed as the GM crop released in North America for commercial cultivation using the technique of:

a) Mutation b) Hybridization

c) Chromosome Doubling d) Gene silencing

39. Biennial bearing in apple is due to:

a) Heavy crop load b) GA3 synthesis in seeds

c) Phenol accumulation d) More fruiting

40. Arka Surya breed up to how many generations:

a) F-14th generation b) F-16th generation

(C) F-17th generation d) F-18th generation

41. For a one hectare area in UHDP system of guava number of plants can be accommodate:

a) 555 b) 1111

c) 2222 d) 4000

42. Oleocellosis is physiological disorder of:

a) Sweet lime b) Grapefruit

c) Mandarin d) Lemon

43. Pink disease in jackfruit is caused by:

a) Botryobasidium b) Colletotricum

c) Rhizopus artocarpi d) Diplodia

44. Somaclonal variation in fruit crop is permitted upto:

a) 1% b) 3-4%

c) 5-6% d) 8-10%

45. Match the following according to their chilling hours requirement:

a) Peach 900-1100

b) Subtropical peach 250-500

c) Asian Pear 500-900

d) Apple 400-600

a) a-iii b-ii c-iv d-i b) a-iii b-ii c-i d-iv

c) a-iii b-i c-iv d-ii d) a-i-b-ii c-iii d-iv

46. Asepsis means?

a) Preservation by high temperature b) Inactivation of Enzymes

c) Keeping out the Microorganisms d) Free from moisture

47. Owens (2008) has divided the grape genotype into how many morphotypes?
    a) 2 b) 3
    c) 4 d) 5
48. In mango stigma receptivity of flowers continues in upto:
    a) 24 hours b) 48 hours
    c) 72 hours or more d) More than 10 days
49. Effective pollination period (days) of litchi varieties in North India:
    a) 4-6 days b) 7-18 days
    c) 20-25 days d) 0-4 days
50. *Psidium cujavalis* guava is having growing habit:
    a) Dwarf b) Vigorous
    c) Semi-dwarf d) Semi vigorous
51. Gisela-5 rootstock of Cherry is:
    a) Vigorous b) Semi vigrous
    c) Semi-dwarf d) Dwarf
52. Black heart and brown spot of pineapple is caused when the storage temperature is below:
    a) 1 °C b) 5 °C
    c) 7 °C d) 8 °C
53. Intense yellow along with purple coloration on leaf is seen due to deficiency of:
    a) Mo b) Mg
    c) P d) Zn
54. In Banana under protected cultivation which physiological disorder affects growth:
    a) Choke throat b) Russeting
    c) Albinism d) Lanky plants
55. Papain quality is measured through:
    a) SPI Test b) Leucine Test
    c) Tyrosine Unit Test d) LEIA test
56. In Banana *M. balbisiana* x *M. acuminata* scoring is?
    a) 26-69 b) 15-25
    c) 70-75 d) 63-69

57. Apple pollinizer recommended for Himachal Pradesh:
    a) Tydeman Early, Worcester, Golden Delicious and Top Red
    b) Tydeman Early, Worcester, Golden Delicious, Red Gold
    c) Yellow Newton, Red gold and Skyline Supreme
    d) Worcester, Tydeman Early, Red Elstar
58. Only variety of mango in which Kesar is used as one of the hybrid parent:
    a) Jawahar b) PKM-1
    c) PKM-2 d) Sai Sugandha
59. System of 3 to 4 lateral branch replacement pruning is practiced in:
    a) Apple b) Kiwifruit
    c) Persimmon d) Passion Fruit
60. In Protected cultivation of strawberry the provided growing liquid medium should have pH:
    a) 6.5 b) 5.5
    c) 5.7 d) 7.0
61. Stanley spot in plum during pre-harvest stage at tree is due to:
    a) High humidity b) Low temperature
    c) High temperature d) Low humidity
62. Most of the Cornell Geneva series rootstock of apple are mainly resistant to:
    a) Codling moth b) Wooly apple aphid
    c) Fire blight disease d) Drought
63. Principle lipids in avocado is:
    a) Oleic and Palmitic b) Oleic and Stearic
    c) Oleic and Linoleic d) Linoleic and Palmitic
64. Super Nova is a mutant variety of:
    a) Almond b) Walnut
    c) Apricot d) Cashew Nut
65. Most productive training system of walnut is:
    a) Open vase system b) Tatura trellis system
    c) Modified Leader System d) Central leader system

66. Genomic constitution of Kunnan variety of banana is:

a)	AAB	b)	AA
c)	AB	d)	AAA

67. Fastrack breeding first time use in which fruit crop?

a)	Peach	b)	Cherry
c)	Plum	d)	Pear

68. Best maturity indices for fruit harvest in pineapple is:

a)	TSS 10° Brix	b)	Specific gravity 0.98 to 1.2
c)	TSS 11° Brix	d)	TSS and Acid 0.1 to 1%

69. Optimum photosynthetic light requirement for papaya cultivation:

a)	1800-1900 µmol m^{-2} s^{-1}	b)	750-1,000 µmol m^{-2} s^{-1}
(C)	1100 µmol m^{-2} s^{-1}	d)	1300 µmol m^{-2} s^{-1}

70. Match the following according to their optimum light saturation point (µmol m^{-2} s^{-1}) requirement

a)	Apple	i)	1300 µmol m^{-2} s^{-1}
b)	Pecan	ii)	1100 µmol m^{-2} s^{-1}
c)	Fig	iii)	1800-1900 µmol m^{-2} s^{-1}
d)	Peach	iv)	700-800 µmol m^{-2} s^{-1}
a)	a-iii b-ii c-iv d-i	b)	a-iii b-ii c-i d-iv
c)	a-iii b-i c-iv d-ii	d)	A-iii, B-iv, C-ii D-i

71. Solar light for better plant growth are desirable where the particles size are in the range of:

a)	10-15 nm	b)	50-100 nm
c)	100-150 nm	d)	200-250 nm

72. Self-Incompatibility in citrus has been decipherable in Xiagunshui lemon first time developed through?

a)	Cisgensis	b)	Invitro mutagenesis
c)	Hybridization	d)	RNA Sequencing

73. Dr. Abhay Dandekar is related to breeding of which fruit crop?

a)	Walnut	b)	Apple
c)	Plum	d)	Peach

74. 1st Indian variety develop through *in ovulo* embryo rescue technique?
    a) Flame Seedless b) Punjab Purple
    c) Pusa Purple Seedless d) Thompson Seedless

75. In apple the bark and bud tolerant to the low temperature by:
    a) Super cooling b) Prevent intra-cellular ice formation
    c) Acclimatization d) Freezing tolerance

76. What is the cross of Vengurla – 6 hybrid of Cashew nut:
    a) Ansur Early x Mysore Kotekar b) Vetore 56 x Ansur-1
    c) Vengurla 3 x M 10 d) BLA-139-1 x K30-1

77. Which of the following grape variety is not recommended to the trained in bower system:
    a) Syrah b) Thompson Seedless
    c) Perlette d) Flame Seedless

78. In summer ovule longevity is shorter following a heavy crop year which may be prolonged in certain cases by application of:
    a) N b) Zn
    c) Fe d) P

79. Polyembryony is first time discovered by:
    a) Leuwenhoek b) Stout
    c) Bergh d) Nirody

80. Stamen's abortion and sub sequent poor fruit set may occur in strawberry when light intensity drops to a level of:
    a) 4400 MW/ m^2 b) 6500 MW/ m^2
    c) 7000 MW/ m^2 d) 8000 MW/m^2

81. The fruit type of pomegranate is:
    a) Balausta b) Hesperidium
    c) Capsule d) Syconus

82. Paclobutrazol is the most popular growth retardant used in production altars:
    a) MAV - pathway b) Photosystem - II
    c) Anaerobic pathway d) Auxin - biosynthesis

83. Gouging is a cultural operation practiced in which fruit crop?
    a) Banana b) Pineapple
    c) Strawberry d) Date palm

84. Prunes botanically belongs to:
    a) *Prunus domestica* b) *Prunus misa*
    c) *Prunus armainiaca* d) *Prunus salicinia*
85. Which of the following avocado variety is developed through hybridization:
    a) Hass b) Pollock
    c) Fuerte d) Waldin
86. Pione-a Tetraploid grape is commercially cultivated in
    a) Spain b) Italy
    c) USA d) Japan
87. Banana multiple crossing variety:
    a) Udhayam b) CO-1
    c) CO-2 d) CO-3
88. The duration of flower initiation to anthesis in mango under tropical condition:
    a) 1 week b) 2 week
    c) 3 week d) 4 week
89. Transgenic F1 papaya is:
    a) T.N. No. 1 b) Rainbow
    c) Sunset Solo d) Sunup
90. Which rootstock of mango enhance resistance to fruit fly by increasing firmness:
    a) Piqueno b) *Mangifera kasturi*
    c) Criollo d) Hilacha
91. Grape rootstock has been evolved using highest no. of donor parent:
    a) 41 B b) 1103 Paulsen
    c) Fercal d) Freedom
92. Biotic stress has been introduced to India through OGV in grape powdery mildew from EU:
    a) GFLY virus b) Viroid's & Phytoplasma
    c) Bacterial Spot d) Fungus (PM)
93. Powdery Mildew Resistant Loci Ren-6 and Ren-7 are located on chromosome no.:
    a) 2 and 3 b) 6 and 7
    c) 14 and 18 d) 9 and 19

94. Case variety of grape has ploidy level:

a) 2X b) 3X
c) 4X d) 6X

95. For colorless in fruits which pigment is responsible ...

a) Anthocyanin b) Flavanols
c) Proanthocyanin d) Carotenoids

96. The International Treaty on Plant Genetics Resources for Food and Agriculture was sign in...:

a) 2002 b) 2005
c) 2007 d) 2020

97. The base temperature for calculation of Growing Degree Days (GDH) in Fruit Crops...

a) 2.2 °C b) 4.4 °C
c) 5.5 °C d) 7 °C

98. Which fruit is rich in pectin:

a) Almond b) Cashew Apple
c) Apple d) Walnut

99. The technology to distinguish zygotic and nucellar polyembryony in citrus species was first time reported using:

a) RFLP b) RAPD
c) Isozymes d) AFLP

100. Notching is an important practice of which fruit crop:

a) Aonla b) Walnut
c) Fig d) Banana

101. Which hormone is use to improve fruit setting:

a) Ethylene b) ABA
c) NAA d) Salicylic acid

102. Arka Neelachal Kesari mango is a clonal selection of which variety:

a) Suwernarekha b) Sindhuri
c) Gulabkhas d) Lal Sindhuari

103. Pectin is present in fruits in the form of:

a) Protopectin b) Pectin
c) Pectic acid d) Calcium pectate

104. Best Degreening temperature and relative humidity for citrus is:

a) 27°C and 85-95% RH
b) 17°C and 85-95% RH
c) 7°C and 85-95% RH
d) 37°C and 85-95% RH

105. Browning of peel followed by internal pulp browning in apple during storage due to:

a) High ethylene
b) Accumulation of hydrocarbons
c) High temperature
d) Enzymatic reaction

106. Cheapest method of preservation is:

a) Canning
b) Sterilization
c) Drying
d) Freezing

107. Dwarfness in Mango is controlled by which gene:

a) Single dominant gene
b) Single recessive gene
c) Poly genes
d) Under cytoplasmic control

108. Inheritance pattern of fruit size in mango is controlled by:

a) Single dominant gene
b) Single recessive gene
c) Polygenes
d) Under cytoplasmic control

109. Prabhat a peach hybrid is a cross between:

a) *Florida Sun × Sharbati*
b) *Sharbati × Florida Sun*
c) *P. persica × P. davididasa*
d) *Sharbati × Shan-e-Punjab*

110. Which of the following is shelf male sterile variety of Peach:

a) July Elberta
b) Halberta
c) J.H. Hale
d) Sun Gold

111. To accommodate 33% pollinizer plants, it's planting in apple should be:

a) Every 2nd Row
b) Every 3rd Row
c) Every 4th Row
d) Every 5th Row

112. Coffee was introduced in India by Buda Budan during:

a) 13th Century
b) 14th Century
c) 16th Century
d) 17th Century

113. Carotene fraction which contains hydroxyl group:

a) Anthocyanin
b) Xanthophyll
c) Lycopene
d) Carotenoid

114. Maximum grapes are preserved by which method by:
   a) Wine making
   b) Canning
   c) Dehydration
   d) By Freezing

115. Temperature in naturally ventilated green house is maintained through:
   a) Side Windows
   b) Vents
   c) by Opening the door
   d) By hydro-coolers

116. In protected cultivation which structural material protect from high wind velocity:
   a) Bracing
   b) Gable
   c) Pole
   d) Vent

117. Is a first hybrid variety of pineapple developed in India:
   a) Kew
   b) Amritha
   c) Mauritius
   d) Boron

118. Enzyme responsible for loss of tissue firmness in fruits during ripening is:
   a) Hydrolyses
   b) Catalyzes
   c) Polygalacturonase
   d) Pectinesterases

119. Whole genome sequencing in pomegranate has been reported in which variety:
   a) Ganesh
   b) Mridula
   c) Bhagwa
   d) Spanish Ruby

120. Flavoring compound in strawberry is:
   a) Methyl esterase
   b) Isoamyl acetate
   c) Ethyl butanoate
   d) 3-hexenal

121. Headquarter of UPOV is situated at:
   a) Rome
   b) Washington, DC
   c) Geneva
   d) Germany

122. Somatic chromosome no. of Illaichi variety of Ber is:
   a) $2x = 24$
   b) $4x = 48$
   c) $6x = 72$
   d) $8x = 96$

123. Miracle fruit of China is:
   a) Litchi
   b) Kinnow
   c) Kiwifruit
   d) Strawberry

124. Which species of kiwifruit is cold tolerant and suitable for interspecific hybridization:
   a) *A. delicious* b) *A. chinensis*
   c) *A. arguata* d) A. kolomikta

125. Compound can be used in *In vitro* ploidy manipulation is:
   a) Oryzalin b) Trifluralin
   c) APM d) DMSO

126. Summer pruning in ber is done in which month:
   a) March b) April
   c) May d) June

127. International Biodiversity Day is celebrated on:
   a) 22 March b) 22 April
   c) 22 May d) 22 June

128. Bio fortified variety of pomegranate Solapur Lal has the parentage of:
   a) Bhagwa x Jyoti b) Ganesh x Nana
   c) Bhagwa x Gulesha Red d) Bhagwa x (Ganesh x Nana) x Daru

129. Technique of obtain viable seed in interspecific hybridization in papaya
   a) Endosperm culture b) Embryo Rescue
   c) *In vitro* mutagenesis d) Haploid production

130. On the basis of respiration rate apple is classified as:
   a) Low b) Very Low
   c) Moderate d) High

131. The nutrient works only act as catalyst:
   a) Cu b) B
   c) Zn d) Mo

132. Largest *invitro* collection of *Musa* germplasm in the world is maintained by:
   a) INIBAP b) IITA
   c) ITC d) CGIAR

133. During summer pruning when temperature is exposed to above 30°C in grapes
   a) Bud differentiation b) Dead-arms
   c) Flowering d) Growth cease

134. Variety of grape which forms inverted bottleneck formation when grafted on Dogridge rootstock:

a) Thompson Seedless b) Perlette
c) Anab-e-Shahi d) Beauty Seedless

135. Ronnse, Kryder, Pomeroy and Large flower are the botanical cultivars of:

a) *Citrus macrophylla* b) *Poncirus trifoliata*
c) *Severinia buxifolia* d) *C. jambhiri*

136. Recently matted row system of planting is commercially followed in:

a) Pineapple b) Kiwifruit
c) Strawberry d) Apple

137. Match the following training systems according to their originated countries

A.	Cordon	i)	France
B.	Meadow	ii)	Israel
C.	Spindle Bush	iii)	Germany
D.	Tatura Trellis	iv)	Australia

a) a-i b-ii c-iii d-iv b) a-iii b-ii c-i d-iv
c) a-iii b-i c-iv d-ii d) A-iii, B-iv, C-ii D-i

138. In Robusta banana paired row planting system (1.5x1.5x2.0m) recorded a total yield of:

a) 43.10 t/ha b) 52 t/ha
c) 58.08 t/ha d) 65.91 tonnes/ha

139. Reflective mulches do improve the light level in the .... portion.

a) Lower canopy b) Middle canopy
c) Inside middle canopy d) Upper canopy

140. Banana species having centre of diversity in eastern India is:

a) *Musa balbisiana* b) *M. acuminate*
c) *M. basjoo* d) *M. ingens*

141. Owens has classified grape into how many morphotypes:

a) Two b) Three
c) Four d) Five

142. In hilly areas fruit cultivation is not possible above the .... slope.

a) 40 ° b) 52°
c) 55° d) 65°

143. Under the protected cultivation tip burning in banana is due to:

a) High temperature
b) Low temperature
c) High humidity
d) Low humidity

144. Match the column:

Fruit crop		Mean temperature for growth	
A.	Apple	i)	20 °C
B.	Mango	ii)	30 °C
C.	Banana	iii)	27 °C
D.	Persimmon	iv)	24 °C

a) a-i b-ii c-iii d-iv
b) a-iii b-ii c-i d-iv
c) a-iii b-i c-iv d-ii
d) A-iii, B-iv, C-ii D-i

145. Embryo rescue in fruit crops first time successfully done by:

a) Hanning
b) Tokey
c) Theophrastus
d) Willey

146. Aroma in over-ripe banana is due to:

a) Eugenol
b) Hexenal
c) Isopentanol
d) Valencene

Answers Key

1.	(b)	2.	(c)	3.	(c)	4.	(a)	5.	(a)	6.	(a)	7.	(c)
8.	(c)	9.	(a)	10.	(a)	11.	(a)	12.	(a)	13.	(b)	14.	(b)
15.	(b)	16.	(c)	17.	(c)	18.	(d)	19.	(a)	20.	(c)	21.	(b)
22.	(b)	23.	(a)	24.	(b)	25.	(c)	26.	(a)	27.	(c)	28.	(c)
29.	(b)	30.	(d)	31.	(b)	32.	(b)	33.	(c)	34.	(b)	35.	(c)
36.	(d)	37.	(a)	38.	(d)	39.	(b)	40.	(a)	41.	(d)	42.	(d)
43.	(a)	44.	(b)	45.	(a)	46.	(c)	47.	(a)	48.	(c)	49.	(b)
50.	(b)	51.	(d)	52.	(c)	53.	(c)	54.	(d)	55.	(c)	56.	(a)
57.	(b)	58.	(d)	59.	(b)	60.	(c)	61.	(c)	62.	(c)	63.	(a)
64.	(a)	65.	(c)	66.	(c)	67.	(c)	68.	(b)	69.	(a)	70.	(d)
71.	(c)	72.	(d)	73.	(a)	74.	(c)	75.	(b)	76.	(b)	77.	(c)
78.	(a)	79.	(a)	80.	(a)	81.	(a)	82.	(a)	83.	(a)	84.	(a)
85.	(c)	86.	(d)	87.	(b)	88.	(d)	89.	(b)	90.	(c)	91.	(c)
92.	(a)	93.	(d)	94.	(c)	95.	(b)	96.	(a)	97.	(d)	98.	(c)
99.	(c)	100.	(c)	101.	(c)	102.	(c)	103.	(d)	104.	(a)	105.	(b)
106.	(c)	107.	(b)	108.	(c)	109.	(b)	110.	(c)	111.	(b)	112.	(c)
113.	(b)	114.	(c)	115.	(b)	116.	(a)	117.	(b)	118.	(c)	119.	(c)
120.	(c)	121.	(c)	122.	(d)	123.	(c)	124.	(d)	125.	(a)	126.	(c)
127.	(c)	128.	(d)	129.	(b)	130.	(a)	131.	(a)	132.	(c)	133.	(b)
134.	(a)	135.	(b)	136.	(c)	137.	(a)	138.	(d)	139.	(a)	140.	(a)
141.	(a)	142.	(a)	143.	(a)	144.	(a)	145.	(b)	146.	(c)		

18

ICAR – NET
Fruit Science Exam (Sample Paper-1)

1. Apple was introduced in India in the year

   a) 1865 b) 1855
   c) 1900 d) 1910

2. To break bud dormancy, most of delicious apple varieties require chilling hours of

   a) 200-300 b) 500-600
   c) 700-800 d) 1000-1600

3. Late season variety of apple is

   a) Granny Smith b) Red Gold
   c) Starking Delicious d) Michal

4. Which is not a spur type variety

   a) Starkrimson b) Red Chief
   c) Ace Spur d) Vance Delicious

5. For cross pollination and good fruit set in apple pollinizers are planted in a proportion of

   a) 5% b) 10%
   c) 15% d) 33%

6. Golden Delicious variety of apple takes how many days from full bloom to harvest

   a) 90 b) 110
   c) 148 d) 188

7. Which is not a serious pest of apple

   a) Woolly apple aphid b) Red mite
   c) Leaf curl aphid d) San Jose Scale.

8. Apple scab is caused by
   a) Fungus b) Nematode
   c) Bacteria d) Virus
9. Which is main factor of low productivity in apple
   a) Nutrition b) Varietal factor
   c) Pathological d) Climatic factor
10. Cankers in apple trees are controlled effectively with the spray of.
   a) Bavistin b) Diathane M-45
   c) Copper OxyChloride d) Dodine
11. Plumcot is hybrid between plum and apricot and comparises of
   a) 50% plum and 25% apricot b) 25% plum and 50% apricot
   c) 50% plum and 50% apricot d) 50% plum and 75% apricot
12. Nematode resistant rootstocks of peach are
   a) GF 667 b) St. Julien
   c) Siberian C d) Shalil
13. Fruit thinning in peaches is done with the spray of
   a) GA b) IBA
   c) Ethephone d) Urea
14. Apricot belongs to sub genous
   a) Amygdalus b) Cerasus
   c) Prunophora d) Prunus
15. Which is drying type of apricot variety
   a) Royal b) New Castle
   c) Halman d) Harcot
16. Recommended Seedling rootstocks for apricot is
   a) Wild apricot b) Peach
   c) Behmi d) Almond
17. Plum with high sugar content is called
   a) Nectarine b) Plumquat
   c) Prune d) None of above

18. Which is not a variety of Japanese plum?

a)	Santa Rosa	b)	Red Beaut
c)	Meriposa	d)	Green Gage

19. Botanical name of Japanese plum is

a)	*Prunus insititia*	b)	*Prunus spinosa*
c)	Prunus salicina	d)	Prunus domestica

20. Plum tree is trained as

a)	Open centre system	b)	Modified centre leader system
c)	Central leader system	d)	Tatura trellis system

21. The origin of almond is

a)	China	b)	America
c)	Europe	d)	Central asian mountain areas.

22 The fat percentage in Kernels of almond is

a)	15	b)	25
c)	54	d)	74

23 Pollination in almond is done by

a)	Wind	b)	Honeybees
c)	Aphid	d)	None of above

24 Clonal rootstock of almond is

a)	GF677	b)	Pixy
c)	Colt	d)	Behmi

25 Kiwifruit is originated in

a)	New Zealand	b)	China
c)	India	d)	Australlia

26 Which is staminate variety of kiwifruit

a)	Hayward	b)	Bruno
c)	Monty	d)	Tomuri

27 Kiwifruit is rich in

a)	Vitamin C	b)	Vitamin A
c)	Protein	d)	Fat

28 At maturity fruit TSS of kiwifruit should be

a) 4.5 Brix b) 6.2 Brix

c) 7.2 Brix d) 8.2 Brix

29 Kiwifruit bear fruit on

a) Spur b) One year shoot

c) Two year shoot d) Current season growth

30 Cultivated strawberry is

a) Hexaploid b) Diploid

c) Triploid d) Octaploid

31 The best time of planting in northern India is

a) July-August b) September-October

c) December-January d) March- April

32 Which is day neutral variety of strawberry

a) Selva b) Chandler

c) Carmarosa d) Sweet Charlie

33 Strawberry is planted at a spacing of

a) 10x10 cm b) 20x20 cm

c) 25x50 cm d) 60x60 cm

34 Basic chromosome (x) number of persimmon is

a) 8 b) 16

c) 24 d) 40

35 Persimmon is national fruit of

a) Indonesia b) China

c) Korea d) Japan

36 Astringency in Persimmon is removed with the dipping of fruits in solution of

a) NAA b) GA

c) Ethephone d) IAA

37 Walnut belongs to the family

a) Rosaceae b) Ebenaeae

c) Rutaceae d) None of above

38 Chilling requirement of walnut is

a) 700-1500 hrs b) 200-400 hrs

c) 300-600 hrs d) 1500-1900 hrs

39 Maximum area and production of walnut in India is in

a) Arunachal Pradesh b) Himachal Pradesh

c) Jammu and Kashmir d) Uttarakhand

40 The origin of pecan nut is

a) Japan b) North America

c) China d) Turkey

41 Which is not a variety of pecan nut

a) Franquette b) Mahan

c) Burkett d) Nellis

42 Pecan is planted at a spacing of

a) 4x4 m b) 5x5 m

c) 6x6 m d) 10x10 m

43 Pecan is commercially propagated through

a) T-budding b) Cutting

c) Tissue culture d) Patch budding

44 Summer pruning is practiced in

a) Apple b) Cherry

c) Peach d) Citrus

45 Which variety of grapes is commercially cultivated in north India

a) Thompson seedless b) Anab-a-Shahi

c) Perlette d) Black Champa

46 State having highest productivity in Guava

a) Uttar Pradesh b) Andhra Pradesh

c) Gujrat d) Maharashtra

47 Type of self incompatibility in laquat

a) Saprophytic b) Gametophytic

c) Both d) None

48 Mangosteen family

a) Guttiferae b) Rutaceae

c) Anacardiaceae d) Arecacae

49 Amrit Sagar is a variety of

a) Banana b) Mango

c) Pineapple d) Custard apple

50 Nendran genotype

a) AAB b) ABB

c) AAA d) BBA

51 The fruit crop which has greater biodiversity in India

a) Peach b) Apricot

c) Olive d) Kiwi

52 Vigorous rootstock of grape

a) Dogridge b) 1613

c) Salt creek d) Temple

53 Inheritance pattern of dwarf ness in Papaya is

a) Recessive b) Dominant

c) Both d) None of these

54 Litchi is propagated by

a) Air Layering b) Cutting

c) Grafting d) Budding

55 In mango, spongy tissue is governed by

a) Recessive genes b) Dominant Genes

c) Both d) None of these

56 Pineapple is mostly packed in

a) CFB b) Wooden crates

c) Polythene lined Gunny bag d) Polythene lined bamboo baskets

57 Ornithophilous fruit crop

a) Banana b) Pineapple

c) Jack d) Sapota

58 Cauliflorous fruiting is observed in

a) Jack fruit b) Mango

c) Papaya d) Guava

59 Transgenic papaya var

a) Rainbow b) Co-3

c) Taiwan d) Pusa majesty

60 BT papaya was developed from

a) USA b) India

c) Japan d) China

61 Carambola is rich in

a) Oxalicacid b) Caricaxanthin

c) Lycopene d) Flavonoids

62 Ber budding time

a) July-August b) Sept-Oct

c) May-June d) Jan-Feb

63 Banana cvr, susceptible to Panama wilt is

a) Rastahli b) Moongli

c) Lady finger d) CO-1

64 Postharvest drop of grape berries from bunch can be controlled by spraying the bunches with

a) 50 ppm NAA b) 60 ppm GA

c) 20 ppm NAA d) 40 ppm GA

65 Pre-harvest drop in fruits is controlled by

a) GA b) BA

c) 2, 4, 5-T d) NAA

66 Auxin precursor is

a) Tryptophan b) Terpenoids

c) Isopentenyl d) None of these

67 The pigment present over the surface of mature orange during harvest is

a) Carotene b) Chlorophyll

c) Naringin d) Anthocyanins

68 Transgraft has

a) Transgenic rootstock b) Transgenic scion

c) Both are transgenic d) None of these

69 First transgenic fruit crop

a) Walnut a) Pear

c) Apple d) Papaya

70 The nutrient element which is immobile ion soil & hence applied only as a basal dose is

a) P b) K

c) N d) Mg

71 The raising problem in protected cultivation is

a) Nematodes b) Fruit fly

c) Fruit Borer d) Aphids

72 Most common type of protected structure used by Indian nursery men is

a) Net house b) Green house

c) Poly house d) None of these

73 The crop grown commercially in protected structures in India is

a) Strawberry b) Papaya

c) Pineapple d) Banana

74 0.1 g of a compound mixed in 10 lit of water gives

a) 1000 ppm b) 100 ppm

c) 10 ppm d) 10000 ppm

75 Dry neck is a disorder in

a) Avocado b) Mango

c) Apple d) Pineapple

76 Ca deficiency is observed in

a) Top leaves b) Lover leaves

c) Bottom leaves d) None of these

77 Mountan papya

a) *C.gracilis* b) *C.papaya*

c) *V.monica* d) *V. candamarcensis*

78 Deciduous citrus spp.

a) Trifoliate orange b) Tangerine orange

c) Tangerine d) None of these

79 Sapota fruit is botanically

a) Berry b) Drope

c) Nut d) None of these

80 Opt. C:N ratio for balance growth & flowering is

a) CCC: NN b) NN: CCC

c) CCC: NNN d) NN: NN

81 Papaya spp having virus resistance source is

a) *C.gracilis* *b) C.papaya*

c) V.monica *d) V. cauliflora*

82 Triple sigmoid growth curve

a) Kiwi fruit b) Mango

c) Apple d) Litchi

83 Single sigmoid growth curve

a) Kiwi fruit b) Mango

c) Apple d) Litchi

84 Highest vitamin C content

a) West Indian Cherry b) Aonla

c) Guava d) Citrus

85 Spring crop in fig is called as

a) Breba b) Ambe bahar

c) Mrig bahar d) Haste Bahar

86 Alternate bearing is not so common in

a) Sapota b) Tamarind

c) Apple d) Citrus

87 Storage temperature for lime is

a) 0-1°C b) 5-6 °C

c) -1°C d) 4-10 °C

88 Granulation disorder is found in

a) Papaya b) Citrus

c) Apple d) Almond

89 In the world, grapes are mostly processed into

a) Raisin b) Juice

c) Wine d) Jam

90 Total no. of AEZ in India

a) 60 b) 50

c) 40 d) 20

91. Match the following

(A)		(B)	
1.	Banana	a)	Day natural plant
2.	Peach	b)	None climatic
3.	Cashew	c)	climatic
4.	Apple	d)	Marginal placentation
5.	Litchi	e)	Mallic acid

a) 1-a.2-c,3-b,4-e,5-d a) 1-d.2-c,3-b,4-e,5-a

c) 1-a.2-b,3-c,4-e,5-d d) 1-e.2-c,3-b,4-a,5-d

92. Match the following

1.	Peach	a)	Bitter pit
2.	Almond	b)	J.H.Hale
3.	Plum	c)	Colt
4.	Apple	d)	Pear less
5.	Cherry	e)	President

a) 1-b,2-c,3-,e,4-a,5-d b) 1-b,2-d,3-,e,4-a,5-c

c) 1-e,2-d,3-,a,4-a,5-c a) 1-a,2-d,3-,e,4-b,5-c

93. Match the following

1.	Sapota	a)	Papain
2.	Mango	b)	Anthocyanin
3.	Papaya	c)	Carotene

4. Ethylene d) Yellow Streak Below Skin
5. Strawberry (e) Methionine

a) 1-a,2-c,3-d,4-e,5-b b) 1-c,2-d,3-a,4-e,5-b
c) 1-d,2-c,3-a,4-e,5-b d) 1-c,2-d3-a,4-b,5-e

94. Match the following

1. Vineyard of india a) Nasik
2. Zygodormany b) Litchi
3. Znsulin c) Pineapple
4. Red rot d) Jammun
5. Yellow disease e) Aonla

a) 1-a,2-e,3-d,4-b,5-c b) 1-a,2-e,3-d,4-b,5-c
c) 1-a,2-e,3-d,4-b,5-c d) 1-a,2-e,3-d,4-b,5-c

95. Match the following

1. Very low a) Strawberry
2. Low b) Mango
3. Moderate c) Snap melon
4. High d) Apple
5. Very high e) Apricot

a) 1-e,2-d,3-a,4-b,5-c b) 1-c,2-d,3-b,4-a,5-e
c) 1-e,2-b,3-d,4-a,5-c d) 1-e,2-d,3-b,4-a,5-c

96. Match the following

1. IBA- 500 ppm a) Sapota
2. Rayan b) Improve berry shape and Size
3. Adriatic c) Rooting of airlayering
4. Sardar d) Fig
5. GA3 e) Guava

a) 1-c,2-a,3-b,4-e,5-d b) 1-c,2-e,3-d,4-a,5-b
c) 1-c,2-a,3-d,4-e,5-b d) 1-c,2-a,3-d,4-b,5-e

97. Match the following

1. Dodridge a) Thompson seedless
2. Khirini b) Basari

3. Rangpurlime — c) Sapota
4. Banana — d) Citrus
5. Grape — e) Grape

a) 1-e,2-c,3-d,4-b,5-a
c) 1-b,2-c,3-d,4-e,5-a
c) 1-a,2-c,3-d,4-b,5-e
d) 1-e,2-d,3-c,4-b,5-a

98. Match the following

1. Sai sarbati — a) Sapota
2. San Pedro — b) Pineapple
3. Cricket Ball — c) Kagzi lime
4. Shahi — d) Fig
5. Mauritius — e) Litchi

a) 1-b,2-d,3-a,4-e,5-c
b) 1-c,2-d,3-a,4-e,5-b
c) 1-c,2-d,3-b,4-e,5-a
d) 1-a,2-d,3-c,4-e,5-b

99. Match the following

1. Mango — a) 30
2. Sapota — b) 48
3. Litchi — c) 40
4. Ber — d) 18
5. Pomegranate — e) 26

a) 1-c, 2-a, 3-e, 4-b, 5-d
b) 1-d, 2-e, 3-a, 4-b, 5-c
c) 1-c, 2-e, 3-a, 4-b, 5-d
d) 1-c, 2-b, 3-a, 4-e, 5-d

100. Match the following

1. Pineapple — a) India
2. Persimmon — b) Tropical America
3. Papaya — c) Brazil
4. Ber — d) China
5. Custard apple — e) Central America

a) 1-c, 2-d, 3-e, 4-a, 5-b
b) 1-c, 2-d, 3-e, 4-a, 5-b
c) 1-c, 2-d, 3-e, 4-a, 5-b
d) 1-c, 2-d, 3-e, 4-a, 5-b

101. Match the following

1.	Water core in apple	a)	Ca
2.	Dieback in Litchi	b)	K
3.	Leaf Bronzing in Litchi	c)	B
4.	Calyx end rot of Grape	d)	Cu
5.	Improper finger filling in Banana	e)	Zn

a) 1-c, 2-d, 3-e, 4-a, 5-b   b) 1-c, 2-d, 3-b, 4-a, 5-e

c) 1-d, 2-c, 3-e, 4-a, 5-b   d) 1-c, 2-d, 3-e, 4-a, 5-b

Answers Key

1.	(a)	2.	(d)	3.	(a)	4.	(d)	5.	(d)	6.	(b)	7.	(c)
8.	(a)	9.	(d)	10.	(c)	11.	(c)	12.	(d)	13.	(c)	14.	(c)
15.	(c)	16.	(a)	17.	(c)	18.	(d)	19.	(c)	20.	(a)	21.	(d)
22.	(c)	23.	(b)	24.	(a)	25.	(b)	26.	(d)	27.	(a)	28.	(b)
29.	(d)	30.	(d)	31.	(b)	32.	(a)	33.	(c)	34.	(b)	35.	(d)
36.	(c)	37.	(d)	38.	(a)	39.	(c)	40.	(b)	41.	(a)	42.	(d)
43.	(d)	44.	(a)	45.	(c)	46.	(c)	47.	(b)	48.	(a)	49.	(a)
50.	(a)	51.	(b)	52.	(a)	53.	(a)	54.	(a)	55.	(a)	56.	(c)
57.	(a)	58.	(a)	59.	(a)	60.	(a)	61.	(a)	62.	(a)	63.	(a)
64.	(a)	65.	(c)	66.	(a)	67.	(a)	68.	(a)	69.	(a)	70.	(a)
71.	(b)	72.	(a)	73.	(a)	74.	(c)	75.	(a)	76.	(a)	77.	(d)
78.	(a)	79.	(a)	80.	(a)	81.	(d)	82.	(a)	83.	(c)	84.	(a)
85.	(a)	86.	(c)	87.	(b)	88.	(b)	89.	(c)	90.	(a)	91.	(a)
92.	(b)	93.	(c)	94.	(a)	95.	(d)	96.	(c)	97.	(a)	98.	(b)
99.	(a)	100.	(c)	101.	(d)								

19

ICAR – NET Fruit Science Exam (Sample Paper-2)

1. Partial ringing of branches above a dormant lateral bud is known as

   a) Ringing b) Notching

   c) Smudging d) Nicking

2. Epigenous germination is found in which of the following fruit crops

   a) Mango b) Tamarind

   c) Cashewnut d) All of them

3. The development of seed portion *viz.* two polar nuclei+ sperm nucleus results into which of the following fruit organ

   a) Seed coat b) Perisperm

   b) Endosperm d) Zygote

4. Which of the following fruit seeds based on storage behavior is classified as recalcitrant seed

   a) Passion fruit b) Datepalm

   c) Jackfruit d) Phalsa

5. Which of the following methods is a renovation/rejuvenation grafting

   a) Tongue grafting b) Buttress grafting

   c) Side-veneer grafting d) Epicotyl grafting

6. Peach scion grafted on plum rootstock results in which type of incompatibility

   a) Partial incompatibility b) Translocalised incompatibility

   c) Localised incompatibility d) None of the above

7. Post set fruit drop is a major problem in commercial cultivation of

   a) Guava b) Litchi

   c) Sapota d) Mango

8. Datepalm is

a) Pod b) Drupe
c) Single seeded berry d) Sorosis

9. Based on photoperiodic response, 'Papaya' is classified under

a) Long-day plant b) Short-day plant
c) Day-neutral plant d) None of the above

10. Which of the following have the lateral fruit buds having inflorescence with leafy shoots in leaf axils?

a) Avocado b) Citrus
c) Guava d) Mango

11. 'Marginal' type of placentation is found in which of the following fruits

a) Banana b) Papaya
c) Citrus d) Litchi

12. Which of the following fruits show non-climacteric respiratory pattern

a) Cherry b) Apple
c) Pear d) Blueberry

13. The orange colour in papaya fruit is due to presence of

a) Anthocyanin b) β-carotene
c) Flavinoids d) Caricaxanthin

14. Persimmon fruit is (ploidy level)

a) Auto-triploid b) Auto-tetraploid
c) Auto-hexaploid d) Auto-octaploid

15. Which of the following is a deciduous citrus species

a) Sweet lime b) Rangpur Lime
c) Tahiti lime d) Trifoliate orange

16. Type of inflorescence found in grapes is

a) Panicle b) Spadix
c) Cymose d) Corymb

17. Pineapple fruit is

a) Sorosis b) Berry
c) Aggregate fruit d) Hespiridium

18. Which of the following is a free stone Mango species
    a) *Mangifera pajang* b) *Mangifera similis*
    c) *Mangifera magnifica* d) *Mangifera rufocostat*
19. 'Mallika' is a hybrid between
    a) Amarapali x Sensation b) Rumani x Neelum
    c) Neelum x Dashehari d) Dashehari x Nelelum
20. Which of the following is a synthetic hybrid of banana
    a) Robusta (AAA) b) Nendran (AAB)
    c) Peyan (AAB) d) Bodles altafort (AAAA)
21. Guava is commonly pollinated by
    a) Honey-bees b) Houseflies
    c) Hand pollination d) Bats
22. World's largest tree borne fruit is
    a) Jackfruit b) Watermelon
    c) Durian d) Papaya
23. 'Avocado' belongs to family
    a) Guttiferae b) Lauraceae
    c) Moraceae d) Bromeliaceae
24. 'Chicken Tongue' is a physiological disorder of
    a) Litchi b) Grape
    c) Loquat d) Egg fruit
25. Which of the following is a 'Pollinizer' variety in apple
    a) Starking delicious b) Red fuji
    c) Golden delicious d) Jonagold
26. Major acid found in pear is
    a) Citric acid b) Malic acid
    c) Tartaric acid d) Ascorbic acid
27. Origin of Japanese plum is
    a) Japan b) South America
    c) China d) West asia

28. Which of the following statement is true
    a) Peaches and Nectarines are same species
    b) Nectarines is a fuzzless Peach with strong flavour and aroma
    c) Nectarines is due to single dominant gene mutation in Peach for fuzziness
    d) All of the above
29. Colouring compound present in cherry is
    a) Anthocyanin b) Xanthonin
    c) Keracyanin chloride d) β-carotene
30. Indian wild strawberry is
    a) *Fragaria chilonensis* b) *Fragaria virginiana*
    c) *Fragaria vesca* d) *Frageria viridis*
31. Leading producer of kiwifruit in world is
    a) China b) New Zealand
    c) Italy d) France
32. Govind, Roopa and Karan are selections of which nut crop in India
    a) Almond b) Pecan nut
    c) Cashew nut d) Walnut
33. Match list-I with list-II and select the correct answer using the codes given below the list

List-I	List-II
A. *Zizyphus mauritiana*	1. Deciduous
B. *Zizyphus jujube*	2. Dwarfing rootstock
C. *Zizyphus numularia*	3. Evergreen

    a) A-2, B-1, C-3 b) A-1, B-2, C-3
    c) A-3, B-1, C-2 d) A-3, B-2, C-1
34. Edible portion in Aonla is
    a) Mesocarp b) Endocarp
    c) Both a & b d) None of the above
35. Match list-I with list-II and select the correct answer using the codes given below the list

List-I	List-II
A. Sitaphal	1. *Annona atemoya*
B. Lakshmanphal	2. *Annona squamosa*

C.	Hanumanphal	3.	*Annona glabra*
D.	Pond apple	4.	*Annona cherimola*
a)	A-1, B-2, C-3, D-4	b)	A-4, B-3, C-2, D-1
c)	A-2, B-4, C-1, D-3	d)	A-2, B-1, C-4, D-3

36. Which of the following statement is incorrect
    a) Seedlessness in pomegranate is due to lack of lignifications of testa
    b) Pomegranate taste is due to citric acid
    c) Mode of pollination in pomegranate is wind
    d) Both self and cross-pollination is observed in pomegranate

37. Banana is propagated by means of sucker which arise from
    a) Underground rhizome b) Underground corm
    c) Stolon d) Pseudostem

38. In India, dates are harvested at which stage of fruit development
    a) Dang b) Doka
    c) Tamar d) Pind

39. 'Caprification' is a type of cross-pollination found in
    a) Bael b) Fig
    c) Phalsa d) Karonda

40. Mirzapuri, Kagzi Gonda, Kagzi Banarasi are the popular varieties of
    a) Bael b) Ber
    c) Jamun d) Carambola

41. Causal organism of 'Sigatoka leaf spot' disease in banana is
    a) Algae b) Bacteria
    c) Fungus d) Virus

42. Green mould rot is a post-harvested disease caused by *Penicillium digitatum* in
    a) Mango b) Banana
    c) Grape d) Citrus

43. Deficiency of calcium in cherry causes
    a) Bitter pit b) Leaf tip burn
    c) Cork spot d) Cracking

44. National Research Centre for Grape (NRCG) is situated at
    a) Nagpur, Maharashtra b) Pune, Maharashtra
    c) Solapur, Maharashtra d) Trichy, Tamilnadu

45. Development of parthenocarpic fruits due to environmental stimulation is known as
    a) Obligatory parthenocarpy b) Facultative parthenocarpy
    c) Vegetative parthenocarpy d) Stimulative parthenocarpy

46. The term 'Stenospermocarpy, was coined by
    a) Stout b) Hayes
    c) Storey d) Shephered

47. Which of the following is responsible for heating effect in green houses
    a) Photosynthetically active radiations (PAR)
    b) Ultra Violet rays
    c) Infra red radiations
    d) Microwaves

48. MOP is also known as
    a) Potassium sulphate b) Potassium chloride
    c) Potassium nitrate d) Di-ammonium phosphate

49. Match list-I with list-II and select the correct answer using the codes given below the list

List-I	List-II
A. Auxins	1. Terpenoids
B. Gibberellins	2. Tryptophan
C. Abscisic acid	3. Methionine
D. Ethylene	4. Mevalonic acid

    a) A-2, B-1, C-4, D-3 b) A-1, B-3, C-2, D-4
    c) A-3, B-2, C-1, D-4 d) A-2, B-1, C-3, D-4

50. 'Bower' system of training is commonly practiced in
    a) Grapes b) Citrus
    c) Plum d) Guava

51. 'Smudging'- practice of smoking the trees in mango is done
    a) To increase the lateral branches and fruit production
    b) To increase the berry size

c) To determine the flowering time

d) To induce the off-season crops

52. The term 'moist-chilling' is synonym to

a) Scarification b) Dormancy

c) Stratification d) Vernalization

53. Rayan or Khirni is used in India as commercial rootstock

a) Guava b) Sapota

c) Ber d) Citrus

54. Which of the following rootstock is suitable for HDP in apple?

a) Crab apple b) MM 106

c) M 27 d) MM 111

55. Manipulation of flowering in mango is done through application of

a) NAA b) 2,4-D

c) Paclobutrazol d) GA_3

56. Preharvest fruit drop in apple can be prevented by application of

a) NAA @ 2-10 ppm b) NAA @ 10 ppm

c) Ethephon @ 10-100 ppm d) GA_3 @ 50 ppm

57. Which of the following fruits possess axillary bearing habit on current season growth

a) Guava b) Citrus

c) Apple d) Litchi

58. Triple sigmoid growth curve is found in

a) Mango b) Grapes

c) Cherry d) Kiwi

59. Strawberry is

a) Slightly tolerant to acid soil b) Moderately tolerant to acid soils

c) Highly tolerant to acid soil d) Tolerant to alkaline soil

60. Salt tolerance of apple is

a) Very low b) Medium

c) High d) Verry high

61. Plant tissue which develops into fruit flesh in fig is

a) Peduncle b) Mesocarp

c) Thalamus d) Pericarp

62. Number of flowers per panicle in mango are in the range of

a) 10-20 b) 100-500

c) 50-100 d) 1000-6000

63. Largest producer state of citrus fruits in India is

a) Andhra Pradesh b) Maharashtra

c) Punjab d) Madhya Pradesh

64. Which of the following statements is true

a) India is the largest producer of citrus fruits in world

b) Polyembryony is not a common feature of most of the citrus fruits

c) Pummelo, Tahiti lime and Citron are monoembryonic citrus species

d) Self and cross incompatibility does not exist in citrus

65. Red wine is prepared from

a) Apple b) Grape

c) Cashew nut d) Malt

66. Sex in papaya is controlled by single gene with following alleles

a) M_1 and M_2 b) M_2 and m

c) M_1 and m d) $M_{1,}$ M_2 and m

67. Commercial method of propagation in Sapota is

a) Tongue Grafting b) Budding

c) Air layering d) Inarching/Approach grafting

68. 'Ratooning' is done in

a) Pomegrante b) Pineapple

c) Papaya d) Persimmon

69. Which fruit is also known as Queen of tropical fruits or Mystery fruit

a) Avocado b) Litchi

c) Mangosteen d) Pineapple

70. Mexican, Guatemalan and West Indian are the popular races of

a) Avocado b) Passion fruit

c) Carambola d) Jackfruit

71. Rambutan and Longan are relative species of

a) Loquat b) Banana

c) Litchi d) Egg fruit

72. Which of the following is a pome fruit
    a) Apple
    b) Loquat
    c) Both a & b
    d) None of the above
73. National fruit of Malaysia and Indonesia known for its aphrodisiacal properties is
    a) Passion fruit
    b) Banana
    c) Persimmon
    d) Durian
74. Apple bowl of India is
    a) Jammu and Kashmir
    b) Himachal Pradesh
    c) Uttrakhand
    d) Arunachal Pradesh
75. Harvesting stage for mango is
    a) Tapka
    b) Neck fall
    c) T-stage
    d) None of the above
76. Commercial storage temperature recommended for apple is
    a) -1 to 4 ^{0}C
    b) 7 to10 ^{0}C
    c) -0.5 to 0 ^{0}C
    d) 11 to 12 ^{0}C
77. Male sterile peach variety is
    a) Florida prince
    b) Kanto-5
    c) J. H. Hale
    d) Shan-e-punjab
78. Ripening fruits emit which type of gas
    a) Ammonia
    b) Methane
    c) Carbon mono-oxide
    d) Ethylene
79. Chemical used for artificial ripening in banana is
    a) Ethephon
    b) Auxin
    c) KNO_3
    d) Calcium carbide
80. Bitter pit in apple is caused due to deficiency of
    a) Boron
    b) Calcium
    c) Iron
    d) Zinc
81. Granulation is a disorder in
    a) Pomegranate
    b) Citrus
    c) Sapota
    d) Mango

82. Which of the following is a low chill pear cultivar
    a) Patharnakh b) Bartlett
    c) Max red Bartlett d) Starking delicious
83. Hydrogen cynamide is used in grape for
    a) Berry thinning b) Bud break
    c) Berry elongation d) Inducing dormancy
84. Tatura trellis training is commonly used in
    a) Peach b) Plum
    c) Pear d) Apple
85. Guava canker is caused by
    a) Virus b) MLO's
    c) Bacteria d) Fungus
86. Which of the following fruit crops has apetalous flower
    a) Litchi b) Mango
    c) Papaya d) Banana
87. Kiwifruit belongs to family
    a) Malphigiaceae b) Bombaceae
    c) Actinidiaceae d) Rosaceae
88. Dwarfness in papaya is controlled by
    a) Recessive genes b) Dominant gene
    c) Co-dominance d) Duplicate gene
89. Hot water treatment of hard coat seeds for coat permeability is called as
    a) Seed priming b) Scarification
    c) Stratification d) None of these
90. Micro-grafting is used to produce plants free from
    a) Virus b) Fungi
    c) Bacteria d) Nematodes
91. Which of the following is growth retardant?
    a) Auxin b) GA
    c) MH d) ABA

92. Astringency in persimmon fruit is removed by

a) Ethylene b) NAA

c) GA d) ABA

93. Which of the following is a non-astringent variety of persimmon

a) Hachiya b) Hyakuma

c) Fuyu d) Triumph

94. Staminate variety of kiwifruit is

a) Abott b) Hayward

c) Bruno d) Tomuri

95. High monsoon humidity is helpful in pollination of

a) Custard apple b) Papaya

c) Aonla d) Ber

96. Seedlessness in seedless guava is due to

a) Polyploidy b) Stenospermocarpy

c) Parthenocarpy d) None of these

97. Out of following which fruit need bahar treatment

a) Litchi b) Guava

c) Ber d) Aonla

98. In grafting, lower part of plant is known as

a) Bud b) Base

c) Stock d) Scion

99. Seedless variety of mango is

a) Langra b) Alphonso

c) Amrapali d) Sindhu

100. Indian Institute of Horticulture Research (IIHR) is situated at

a) New Delhi b) Bangalore

c) Shimla d) Kolkata

101. Match list-I with list-II and select the correct answer using the codes given below the list

List-I	List-II
A. Banana	1. Bacterial canker
B. Citrus	2. Black tip

C. Mango — 3. Bunchy top

D. Papaya — 4. Ringspot virus

a) A-3, B-1, C-2, D-4 b) A-4, B-3, C-2, D-1

c) A-2, B-4, C-1, D-3 d) A-2, B-1, C-4, D-3

102. Match list-I with list-II and select the correct answer using the codes given below the list

List-I	List-II
A. Mango	1. Rosaceae
B. Guava	2. Bromelliaceae
C. Apple	3. Anacardeaceae
D. Pineapple	4. Myrtaceae

a) A-1, B-2, C-3, D-4 b) A-4, B-3, C-2, D-1

c) A-3, B-4, C-1, D-2 d) A-3, B-1, C-4, D-2

103. Match list-I with list-II and select the correct answer using the codes given below the list

List-I	List-II
A. Mango	1. Seed
B. Aonla	2. Air Layering
C. Papaya	3. Budding
D. Jackfruit	4. Veneer Grafting

a) A-1, B-2, C-3, D-4 b) A-4, B-3, C-2, D-1

c) A-2, B-4, C-1, D-3 d) A-4, B-3, C-1, D-2

104. Match list-I with list-II and select the correct answer using the codes given below the list

List-I	List-II
A. Mango	1. Poor fruit set
B. Grape	2. Malformation
C. Pineapple	3. Fasciation
D. Custard apple	4. Powdery mildew

a) A-1, B-2, C-3, D-4 b) A-4, B-3, C-2, D-1

c) A-2, B-4, C-3, D-1 d) A-2, B-1, C-4, D-3

105. Match list-I with list-II and select the correct answer using the codes given below the list

List-I		List-II	
A.	Bitter pit in apple	1.	Zinc
B.	Little leaf in mango	2.	Calcium
C.	Exanthema in citrus	3.	Boron
D.	Fruit necrosis in aonla	4.	Copper

a) A-1, B-2, C-3, D-4 b) A-4, B-3, C-2, D-1
c) A-2, B-4, C-1, D-3 d) A-2, B-1, C-4, D-3

106. Match list-I with list-II and select the correct answer using the codes given below the list

List-I		List-II	
A.	Apple	1.	King of temperate fruits
B.	Banana	2.	Tree of Paradise
C.	Carambola	3.	Star fruit
D.	Date palm	4.	Head in fire & foot in water crop

a) A-1, B-2, C-3, D-4 b) A-4, B-3, C-2, D-1
c) A-2, B-4, C-1, D-3 d) A-2, B-1, C-4, D-3

107. Match list-I with list-II and select the correct answer using the codes given below the list

List-I		List-II	
A.	Protein	1.	Aonla
B.	Carbohydrate	2.	Almond
C.	Ascorbic acid	3.	Banana
D.	Iron	4.	Datepalm

a) A-1, B-2, C-3, D-4 b) A-4, B-3, C-2, D-1
c) A-2, B-4, C-1, D-3 d) A-2, B-1, C-4, D-3

108. Match list-I with list-II and select the correct answer using the codes given below the list

List-I		List-II	
A.	Balausta	1.	Custard apple
B.	Pepo	2.	Citrus
C.	Hesperidium	3.	Watermelon

D. Etaerio of berries — 4. Pomegranate

a) A-1, B-2, C-3, D-4 — b) A-4, B-3, C-2, D-1

c) A-2, B-4, C-1, D-3 — d) A-2, B-1, C-4, D-3

109. Match list-I with list-II and select the correct answer using the codes given below the list

List-I (Fruit)	List-II (List of origin)
A. Banana	1. China
B. Kiwifruit	2. Central America
C. Avocado	3. Indo-malayan
D. Pomegranate	4. Iran

a) A-1, B-2, C-3, D-4 — b) A-4, B-3, C-2, D-1

c) A-3, B-1, C-2, D-4 — d) A-2, B-1, C-4, D-3

110. Match list-I with list-II and select the correct answer using the codes given below the list

List-I	List-II
A. Rhizome	1. Pineapple
B. Slips	2. Banana
C. Runner	3. Datepalm
D. Offshoot	4. Strawberry

a) A-1, B-2, C-3, D-4 — b) A-4, B-3, C-2, D-1

c) A-2, B-4, C-1, D-3 — d) A-2, B-1, C-4, D-3

111. Match list-I with list-II and select the correct answer using the codes given below the list

List-I	List-II
A. Central Leader	1. Peach
B. Open centre	2. Pomegranate
C. Espalier	3. Walnut
D. Multiple stem system	4. Apple

a) A-1, B-2, C-3, D-4 — b) A-4, B-3, C-2, D-1

c) A-3, B-1, C-4, D-2 — d) A-2, B-1, C-4, D-3

Answers Key

1.	(d)	2.	(d)	3.	(c)	4.	(c)	5.	(b)	6.	(b)	7.	(d)
8.	(c)	9.	(c)	10.	(a)	11.	(d)	12.	(a)	13.	(d)	14.	(c)
15.	(d)	16.	(a)	17.	(a)	18.	(b)	19.	(c)	20.	(d)	21.	(a)
22.	(a)	23.	(b)	24.	(a)	25.	(c)	26.	(b)	27.	(c)	28.	(d)
29.	(c)	30.	(c)	31.	(c)	32.	(d)	33.	(c)	34.	(c)	35.	(d)
36	(c)	37.	(a)	38.	(b)	39.	(b)	40.	(a)	41.	(c)	42.	(d)
43	(d)	44.	(b)	45.	(b)	46.	(a)	47.	(c)	48.	(b)	49.	(a)
50.	(a)	51.	(d)	52.	(c)	53.	(b)	54.	(c)	55.	(c)	56.	(b)
57.	(a)	58.	(d)	59.	(c)	60.	(a)	61.	(a)	62.	(d)	63.	(a)
64.	(c)	65.	(b)	66.	(d)	67.	(d)	68.	(b)	69.	(c)	70.	(a)
71.	(c)	72.	(c)	73.	(d)	74.	(b)	75.	(a)	76.	(a)	77.	(c)
78.	(d)	79.	(d)	80.	(b)	81.	(b)	82.	(a)	83.	(b)	84.	(a)
85.	(d)	86.	(a)	87.	(c)	88.	(a)	89.	(b)	90.	(a)	91.	(d)
92.	(c)	93.	(c)	94.	(d)	95.	(a)	96.	(a)	97.	(b)	98.	(c)
99.	(d)	100.	(b)	101.	(a)	102.	(c)	103.	(d)	104.	(c)	105.	(d)
106.	(a)	107.	(c)	108.	(b)	109.	(c)	110.	(d)	111.	(c)		

20

ICAR – NET
Fruit Science Exam (Sample Paper-3)

1. Low chill varieties of peach needs chilling of
   - a) 50-100 hr
   - b) 250-300 hr
   - c) 600-650 hr
   - d) 800-900 hr
2. Nectarine is a mutant of
   - a) Peach
   - b) Plum
   - c) Apricot
   - d) Almond
3. Nematode resistant rootstocks of peach is
   - a) GF 667
   - b) St. Julien
   - c) Siberian C
   - d) Shalil
4. Fruit thinning in peaches is done with the spray of
   - a) GA
   - b) IBA
   - c) Ethephone
   - d) Urea
5. While pruning of peach tree one year shoots are headed back to
   - a) 25%
   - b) 50%
   - c) 75%
   - d) 90%
6. Plum with high sugar content is called
   - a) Nectarine
   - b) Plumquat
   - c) Prune
   - d) None of above
7. Which is not a variety of Japanese plum?
   - a) Santa Rosa
   - b) Red Beaut
   - c) Meriposa
   - d) Green Gage
8. The best orchard soil management system in plum orchard is
   - a) Clean cultivation
   - b) Cultivation of cover crops
   - c) Mulching
   - d) Sod culture and mulching in tree basin

9. Botanical name of Japanese plum is
   a) *Prunus insititia* b) *Prunus spinosa*
   c) *Prunus salicina* d) *Prunus domestica*
10. Plum tree is trained with
   a) Open Centre System b) Modified Centre leader system
   c) Central leader system d) Tatura trellis system
11. Apricot belongs to Sub-genus
   a) Prunophora b) Amygdalus
   c) Cerasus d) Prunocerasus
12. Which is drying type of apricot variety?
   a) Royal b) New Castle
   c) Halman d) Harcot
13. Apricot is planted at a spacing of
   a) 10x10 m b) 8x8 m
   c) 4x4 m d) 6x6 m
14. The NPK requirement for fully grown apricot tree is.
   a) 250:250:250 g per year b) 500:250:500 g per year
   c) 700:250:700 g per year d) 500:250:700 g per year
15. Recommended seedling rootstocks for apricot is
   a) Wild apricot b) Peach
   c) Behmi d) Almond
16. Other name of apricot is
   a) Zardalu b) Chuli
   c) Coppery fruit d) China miracle fruit
17. Yellow colour of peaches is due to
   a) Anthocyonin b) Carotenoids
   c) Xanthophylls d) None
18. Seeds of peach contain
   a) Amygdalin b) Prunacin
   c) Glycosides d) None

19. The ploidy level of European plum is
    a) Diploid b) Triploid
    c) Tetraploid d) Hexaploid
20. The Plumcot is cross between
    a) Apricot and Plum b) Pulm and Apricot
    c) Peach and Apricot d) Plum and Almond
21. EMA is cultivar of
    a) Apricot b) Plum
    c) Peach d) None
22. Dwarfing rootstock of plum is
    a) Mariana b) Myarobalan B
    c) Wild apricot d) Pixy
23. Male sterile variety of peach is or are
    a) Halberta b) J H hale
    c) Froridasun d) a and b both
24. Chilling requirement of European plum is
    a) 300-600 b) 1000-1200
    b) 700-1000 d) 400-500
25. Suture red spot in peaches is due to
    a) Low Mn b) High F
    c) Low Ca d) Low Zn
26. Chaubattia Alankar and Chaubattia Madhu are cultivars of
    a) Apricot b) Peach
    c) Plum d) Apple
27. Which of the following fruit crop is suitable for cold- arid zone
    a) Peach b) Apricot
    c) Plum d) Nectarine
28. Which of following is best drying type cultivar of apricot
    a) Kaisha b) Halman
    c) Charmagaz d) New Castle

29. Zardalu is wild rootstock of
    a) Apricot
    b) Plum
    c) Peach
    d) Apple
30. Zardalu is native to
    a) India
    b) Japan
    c) China
    d) Brazil
31. Apricot is trained to
    a) Open centre system
    b) Central leader
    c) Modified Centre leader
    d) None
32. Apoplexy is realted to
    a) Apricot
    b) Plum
    c) Avacado
    d) Pear
33. Spur life of Apricot is
    a) 4-5 years
    b) 5-6 years Central leader
    c) 3-4 years Modified
    d) Less than 3 years
34. Edible portion of apricot is
    a) Mesocarp
    b) Mesocarp and endocarp
    c) Pericarp
    d) All of the above
35. White seed kernel variety of apricot is
    a) Khante
    b) Charmagaz
    c) Shakarpara
    d) Rackchekarpo
36. Other name of apricot is
    a) Zardalu
    b) Chuli
    c) Coppery fruit
    d) China miracle fruit
37. Peach belongs to family
    a) Rosaceae
    b) Actinidiaceae
    c) Rutaceae
    d) Apocyanaceae
38. Yellow colour of peaches is due to
    a) Anthocyonin
    b) Carotenoids
    c) Xanthophylls
    d) None

39. Seeds of peach contain

a) Amygdalin b) Prunacin

c) Glycosides d) None

40. Male sterile variety of peach is or are

a) Halberta b) J H hale

c) Froridasun d) a and b both

41. Peach require

a) Moderate level of pruning b) Heavy pruning

c) Light pruning d) None

42. In Tutra trellis system of training in Peach spacing required is

a) 5x1m b) 3x3m

c) 2x1m d) None

43. Maximum number of plants are occupied in which system

i) Tutra trellis ii) Meadow system

iii) HDP iv) Square system

44. Which of the following is dwarf peach variety

a) Australlian Dwarf b) Flordasun

c) J H Hale d) July Elberta

45. Suture red spot in peaches is due to

a) Low Mn b) High F

c) Low Ca d) Low Zn

46. Nectarines are originated from fuzzy skinned peaches as a result of mutation. Mutation is due to

a) Single dominant genes b) Recesssive genes

c) Multiple genes d) None of Above

47. Drying type of plum group is

a) Plumcot b) Aprium

c) Prunes d) None

48. The ploidy level of the European plum is

a) Diploid b) Triploid

c) Tetraploid d) Hexaploid

49. The Plumcot is cross between

a) Apricot and Plum b) Pulm and Apricot

c) Peach and Apricot d) Plum and Almond

50. EMA is cultivar of

a) Apricot b) Plum

c) Peach d) None

51. Dwarfing rootstock of plum is

a) Mariana b) Myarobalan B

c) Wild apricot d) Pixy

52. Fruit thinning agent in plum is

a) DNOC b) Ethaphon

c) 3-CPA d) All the above

53. Chilling requirement of European plum is

a) 300-600 b) 1000-1200

c) 700-1000 d) 400-500

54. Level of fruit thinning required in plum is

a) Less than 20 % b) 25%

c) 25-40% d) More than 40%

55. Low chill variety of plum is

a) Mariposa b) Santa Rosa

c) Beauty d) Titron

56. The seeds of which of the following fruits loose viability very fast after extraction from the fruit

a) Peach b) Apple

c) Plum d) Citrus

57. Vermiculite is a hydrated

a) Mg-Al iron silicate b) Mn-Al iron silicate

c) Ca -Al iron silicate d) Zinc-Al iron silicate

58. The sowing depth of seed is generally

a) 5-6 times of its size b) 3-4 times of its size

c) 1-2 times of its size d) 6-8 times of its size

59. True to type mother plants are maintained in bud wood bank for the supply of

    a) Scion wood b) Seed
    c) Rootstock d) Interstock

60. For raising nursery of evergreen plants the soil should be

    a) Sandy loam b) Clay loam
    c) Sandy clay loam d) None of these

61. Which of the following auxins are commercially used as rooting hormone

    a) 2,4-D b) NAA
    c) IAA d) IBA

62. Apomixes in citrus of the type

    a) Polyembryony b) Recurrent
    c) Vegetative d) Non-recurrent

63. Wooly aphid resistant root stock of apple is

    a) M-27 b) M-9
    c) Crab-apple d) Northern spy

64. The commercial method of grape propagation is

    a) Hard wood cutting b) Seed
    c) Layering d) None of the above

65. Lemon is commercially propagated by

    a) Seed b) Cutting
    c) Budding d) None of the above

66. Litchi is propagated through

    a) Tongue grafting b) Budding
    c) Air layering d) Inarching/Approach grafting

67. Micro-grafting is used to produce plants free from

    a) Virus b) Fungi
    c) Bacteria d) Nematodes

68. In grafting, lower part of plant is known as

    a) Bud b) Base
    c) Stock d) Scion

69. Budding time of ber is
    a) July-August b) Sept-Oct
    c) May-June d) Jan-Feb
70. Most common type of protected structure used by Indian nursery men is
    a) Net house b) Green house
    c) Poly house d) None of these
71. The fruit crop grown commercially in protected structures in India is
    a) Strawberry b) Papaya
    c) Pineapple d) Banana
72. In grafting, lower part of plant is known as
    a) Bud b) Base
    c) Stock d) Scion
73. Pecan is commercially propagated through
    a) T-budding b) Cutting
    c) Tissue culture d) Patch budding
74. Banana is commercially propagated by
    a) Crown b) Sword suckers
    c) Budding d) Water suckers
75. Dwarfing rootstock of mango is
    a) Olour b) Totapuri red small
    c) Alphanso d) Dashehri
76. Hot water treatment of hard coat seeds for coat permeability is called as
    a) Seed priming b) Scarification
    c) Stratification d) None of these
77. Micro-grafting is used to produce plants free from
    a) Virus b) Fungi
    c) Bacteria d) Nematodes
78. Commercial method of propagation in Sapota is
    a) Tongue Grafting b) Budding
    c) Air layering d) Inarching/Approach grafting

79. Banana is propagated by means of sucker which arise from
    a) Underground rhizome
    b) Underground corm
    c) Stolon
    d) Pseudostem
80. Peach scion grafted on plum rootstock results in which type of incompatibility
    a) Partial incompatibility
    b) Trans-localised incompatibility
    c) Localised incompatibility
    d) None of the above

Answers Key

1.	(b)	2.	(a)	3.	(d)	4.	(c)	5.	(c)	6.	(c)	7.	(d)
8.	(d)	9.	(c)	10.	(a)	11.	(a)	12.	(c)	13.	(d)	14.	(d)
15.	(a)	16.	(c)	17.	(a)	18.	(c)	19.	(d)	20.	(b)	21.	(a)
22.	(d)	23.	(d)	24.	(b)	25.	(b)	26.	(a)	27.	(b)	28.	(b)
29.	(a)	30.	(a)	31.	(c)	32.	(a)	33.	(c)	34.	(a)	35.	(d)
36.	(c)	37.	(a)	38.	(a)	39.	(d)	40.	(d)	41.	(b)	42.	(a)
43.	(b)	44.	(a)	45.	(b)	46.	(b)	47.	(c)	48.	(d)	49.	(b)
50.	(a)	51.	(d)	52.	(d)	53.	(b)	54.	(c)	55.	(d)	56.	(d)
57.	(a)	58.	(b)	59.	(a)	60.	(b)	61.	(d)	62.	(a)	63.	(d)
64.	(a)	65.	(a)	66.	(c)	67.	(a)	68.	(c)	69.	(a)	70.	(a)
71.	(a)	72.	(c)	73.	(d)	74.	(b)	75.	(b)	76.	(b)	77.	(a)
78.	(d)	79.	(a)	80.	(b)